T0192001

International Perspectives on Aging

Volume 9

Series Editors

Jason L. Powell
Sheying Chen

For further volumes:
http://www.springer.com/series/8818

Andrew Sixsmith • Gloria Gutman
Editors

Technologies for Active Aging

 Springer

Editors
Andrew Sixsmith
Gerontology Research Centre
Simon Fraser University
Vancouver, BC, Canada

Gloria Gutman
Gerontology Research Centre
Simon Fraser University
Vancouver, BC, Canada

ISBN 978-1-4899-9911-5   ISBN 978-1-4419-8348-0 (eBook)
DOI 10.1007/978-1-4419-8348-0
Springer New York Heidelberg Dordrecht London

Printed on acid-free paper

Springer is part of Springer Science+Business Media (www.springer.com)

Acknowledgments

The editors would like to acknowledge the major contribution of Robert Beringer to the production of this book. We would also like to thank the authors of the chapters for their hard work and commitment.

Contents

Contributors

Arlene Astell CATCH (Centre for Assistive Technology and Connected Healthcare), School of Health and Related Research, University of Sheffield, Sheffield, UK

Elena Avatangelou Research Department, EXODUS S.A., Athens, Greece

Scott Beach University Center for Social and Urban Research, University of Pittsburgh, Pittsburg, PA, USA

Robert Beringer Vibrant Living and Lifestyle Systems INC., Salt Spring Island, BC, Canada

Ilse Bierhoff Stichting Smart Homes, Eindhoven, Eersel, The Netherlands

Paula Byrne Division Primary Care, University Liverpool, Liverpool, Merseyside, UK

Maria Carrillo Medical & Scientific Relations, Alzheimer's Association, National Office, Chicago, IL, USA

Neil Charness William G. Chase Professor of Psychology, Florida State University, Tallahassee, FL, USA

Sara Czaja University of Miami Miller School of Medicine, Miami, FL, USA

Leela Damodaran Information Technology and Social Research Group, Department of Information Science, Loughborough University, Loughborough, UK

Sarah Delaney Work Research Centre, Kimmage, Dublin, Ireland

Vesna Dolničar Faculty of Social Sciences, University of Ljubljana, Ljubljana, Slovenia

Andrew Eccles School of Applied Social Sciences, University of Strathclyde, Glasgow, Scotland

Rory Fisher Division of Geriatric Medicine, Department of Medicine, Sunnybrook Health Science Centre, University of Toronto, Toronto, ON, Canada

Kelly Fitzgerald Western Kentucky University, Affoltern am Albis, Switzerland

Maggie Gibson Veterans Care Program, Parkwood Hospital, St. Joseph's Health Care London, London, ON, Canada

Mary Gilhooly Institute for Ageing Studies, Brunel University, Uxbridge, Middlesex, UK

Gloria Gutman Gerontology Research Centre, Simon Fraser University, Harbour Centre, Vancouver, BC, Canada

Irene Hardill Department of Social Sciences, Northumbria University, Lipman Building, Newcastle, UK

Sandra Hirst Brenda Strafford Centre for Excellence in Gerontological Nursing, University of Calgary, Calgary, AB, Canada

Claire Huijnen Stichting Smart Homes, Duizelseweg, Eersel, Postbus, Eindhoven, The Netherlands

Eunju Hwang Apparel, Housing, and Resource Management, Virginia Tech, Blacksburg, VA, USA

Jeffrey Jutai Interdisciplinary School of Health Sciences, University of Ottawa, Ottawa, ON, Canada

Peter Lansley KT-EQUAL – Knowledge Transfer for Extending Quality Life, School of Construction Management and Engineering, University of Reading, Reading, UK

Babis Magoutas Institute of Communication and Computer Systems, National Technical University of Athens, Zografou Campus, Athens, Greece

Gail Mountain Rehabilitation and Assistive Technologies Group, School of Health and Related Research, University of Sheffield, Sheffield, UK

Sonja Müller empirica Gesellschaft für Kommunikations- und Technologieforschung mbH, Bonn, Germany

Wendy Olphert Information Technology and Social Research Group, Department of Information Science, Loughborough University, Loughborough, UK

Andrew Park Computing Science, Thompson Rivers University, Kamloops, BC, Canada

David Phillips Lingnan University, Lingnan, Hong Kong

Robert Roush Department of Medicine, Baylor College of Medicine, Huffington Center on Aging, Houston, TX, USA

Richard Schulz University Center for Social and Urban Research, University of Pittsburgh, Pittsburg, PA, USA

Andrew Sixsmith Gerontology Research Centre, Simon Fraser University, Harbour Centre, Vancouver, BC, Canada

Kenneth Southall Institut Raymond-Dewar Centre de recherche interdisciplinaire en réadaptation, Montréal, PQ, Canada

Sandra Schoenrade-Sproll Fraunhofer IAO, Nobelstraße, Stuttgart, Germany

Yiannis Verginadis Institute of Communication and Computer Systems, National Technical University of Athens, Zografou Campus, Athens, Greece

Ryan Woolrych Gerontology Research Centre, Simon Fraser University, Harbour Centre, Vancouver, BC, Canada

Chapter 1
Introduction

Andrew Sixsmith and Gloria Gutman

The challenge of population aging requires innovative approaches to meet the needs of increasing numbers of older people. Emerging information and communication technologies (ICTs), such as *pervasive computing* and *ambient assistive living*, have considerable potential for enhancing the quality of life of many older people by providing additional safety and security, supporting mobility, independent living, and social participation. This book presents a snapshot of the current state of the art and serves as a pointer to directions for future research and emerging technologies, products, and services.

The idea for the book originated at the International Society for Gerontechnology's 7th World Conference held in Vancouver, Canada, May 27–30, 2010. Globally, this series of conferences is the foremost platform on the topic of technology and aging; it brings together internationally recognized research leaders from around the world to speak about new developments in R&D in their respective jurisdictions as well as collaborative work across borders and boundaries. The aim was not to produce a book of conference proceedings but rather to invite chapters some of which would be based on presentations in selected symposia at the conference and others specially written for the book so as to reflect the diversity and the most theoretically and technically challenging directions in the field. The resulting book explores these and some of the key application areas.

Rather than presenting specific technical (engineering and computing) problems and solutions, the book is written so as to be accessible to the nontechnical reader. Underpinning the chapters is a gerontological perspective that critically explores the needs of older people in society and their implications for design and technology solutions.

A. Sixsmith, Ph.D. (✉) • G. Gutman, Ph.D.
Gerontology Research Centre, Simon Fraser University, Harbour Centre,
#2800-515 West Hastings Street, Vancouver, BC, Canada V6B 5K3
e-mail: Sixsmith@sfu.ca; gutman@sfu.ca

A. Sixsmith and G. Gutman (eds.), *Technologies for Active Aging*,
International Perspectives on Aging, DOI 10.1007/978-1-4419-8348-0_1,
© Springer Science+Business Media New York 2013

As the title suggests, a key concept behind the book is that of *active aging* which currently occupies a central position in the international discourse on policy for aging, particularly in the 27 European Union countries which declared 2012 the European Year for Active Aging and Solidarity Between Generations (YA2012).

1.1 Active Aging and Gerontechnology: 2002 and Now

The idea of active aging first became prominent among gerontologists in 2002 with the publication of *Active Aging: A Policy Framework*, a contribution of the World Health Organization to the Second United Nations World Assembly on Aging held in Madrid, Spain. This landmark document drew attention to the fact that population aging was the product of two converging trends: more and more people living to be old at the same time dramatic decreases were occurring in fertility rates; that population aging was to occur in both the developed and the developing worlds; and that if it was to be a positive experience for countries and individuals, "longer life must be accompanied by continuing opportunities for health, participation and security" (WHO, 2002, p. 12). *Active aging* was identified as the process for achieving this vision. This term, adopted by WHO in the late 1990s, was meant to convey a more inclusive message than healthy aging—in particular, to draw attention to the fact that health care was only one of a number of determinants of how people and populations age (Kalache & Kickbusch, 1997). It was also meant to include persons with less than optimum physical and/or mental health. The latter is clearly reflected in the definition of *active* provided in the policy framework document:

> The word "active" refers to continuing participation in social, economic, cultural, spiritual and civic affairs, not just the ability to be physically active or to participate in the labour force. Older people who retire from work and those who are ill or live with disabilities can remain active contributors to their families, peers, communities and nations. Active aging aims to extend healthy life expectancy and quality of life for all people as they age, including those who are frail, disabled and in need of care (WHO, 2002, p. 12).

While the intention was laudatory, one of the unintended consequences of this definition is that since 2002 much of the effort by individuals working in the field of aging, and especially in the subfield of gerontechnology, has been to concentrate on providing assistive technologies and devices for the frail and disabled. While this is an important line of research, in this book, our intention is to explore how technology can positively contribute to the health and quality of life of *all* seniors and to put the spotlight on seniors as proactive participants in a digital society. These critical ideas are developed in Chap. 2 of this volume where it is argued that the research and development agenda needs to expand and address the needs of several key subgroups within the seniors' market including healthy and active seniors, people with chronic diseases, people with dementia, and people with mild cognitive impairment. In addition, more effective translation of existing research knowledge is needed if we are to develop useful new products and services that will, in fact, be accepted and used by seniors.

In the chapters that follow Chap. 2, we have attempted to highlight gerontechnology developed for the four groups identified above. We have also attempted to identify other gaps in the knowledge base. For example, Chap. 3 looks at ethical issues which attend the use of assistive technologies. Three areas are examined: first, ethical approaches commonly in use and their limitations for application in the field of assistive technologies; second, the ethical issues which arise around the design and execution of research with users of assistive technologies; and third, the contribution of assistive technologies to the search for that mainstay of moral philosophy, *the good life*. The latter is broached through a discussion of quality of life indicators.

In Chap. 4 the focus is on the use and acceptance of technology by seniors. The chapter begins with a description of e-health applications that facilitate self-care, in particular online health information seeking. Interesting data are presented on the extent to which seniors search the Internet for health information and what they search for and important questions are raised as to how they deal with the vast amount of information that is available. Discussion then turns to monitoring technologies that may permit early detection of illness—with the goal of enabling preventive measures to be put in place that will delay or deter the onset of disability. A key issue here is privacy trade-off and what information people are willing to share about themselves and their day-to-day functioning and with whom.

Chapter 5 introduces the concept of resilience which can be understood as the capacity to adapt to and recover from adversity. The context that it is applied to is natural and human-made disasters which are a source of large-scale adversity that disproportionately impacts older adults. The authors examine the potential for technology to promote disaster resilience for older adults. Technological solutions that could be applied to practical problems are considered for each phase of the emergency management cycle (prevention/mitigation, preparedness, response, recovery). The technologies considered include tracking and mapping systems, intelligent building systems, medical and assistive devices, communication and notification systems, needs assessment strategies, medical support strategies, security strategies, and reconstruction strategies. This chapter concludes with a discussion of factors that may influence the acceptance, uptake, and application of technology to increase disaster resilience in this population.

The Active Aging Policy Framework requires action on three basic pillars: health, security, and participation. Assistive technology provides a platform to support participation, but some devices work better than others and/or even within a particular disability group, are more effective for some users than for others. While there has been significant progress in developing measures of the effectiveness of assistive technology, some important challenges remain because there are many things that contribute to a person's level of participation. Chapter 6 reports the experience of developing the Assistive Technology Outcomes Profile for Mobility (ATOP/M) and lessons learned for measuring the participation of older individuals who use mobility assistive technology.

The focus of Chap. 7 is on technology developed for persons with dementia and their carers. This chapter outlines the services that are commonly provided following diagnosis including how assistive technologies and in particular devices that are categorized as telecare are increasingly becoming part of the *service offer*. Some of the benefits and potential issues with telecare are posed. This is followed by illustrations of how new technology can be developed to meet the changing needs of people with dementia in a wide range of domains including recreation and leisure. This section includes examples of technology developed in partnership with people with dementia in a way that maximizes their retained abilities while playing down the aspects of functioning they find difficult. It then goes on to examine the potential of using everyday technologies to assist people with dementia. Finally, it looks towards a future where the full potential of technology is harnessed to enhance quality of life for people at all stages of the dementia trajectory.

Chapter 8 describes a European research project, SOPRANO (Service-oriented Programmable Smart Environments for Older Europeans), which developed supportive environments for older people based on the concept of Ambient Assisted Living (AAL). AAL uses pervasive information and communication technologies (ICTs) to enable older persons to live independently in their own homes and to increase their quality of life. The focus of this chapter is on the steps that need to be taken to meet these objectives, starting from a user-centered approach to define the AAL system and ending with the actual implementation of developed technology and services in real-life situations.

Computer-generated graphics are becoming more advanced and sophisticated, and realistic simulations of the built environment offer a useful tool to researchers and designers. Potential changes to the built environment to create more age-friendly settings can be tested using a simulation before actual design changes are implemented. People are able to experience the virtual environment and provide their opinions, and researchers can observe their interactions in a safe and secure context. Chapter 9 discusses these and other practical applications of a computer-generated virtual environment as a tool for age-friendly design visualization and communication.

Chapter 10 explores the importance of happiness in older adult's lives and the contribution technology can make to people achieving it. In this context, happiness can be seen as shorthand for the positives in life, the things that make life meaningful and fulfilling. Besides health and well-being, these include satisfying relationships, opportunities for continued achievement, as well as hope and optimism for the future. Building on the foundations of positive psychology, this chapter examines ways in which technology can contribute to making the lives of older people happier and ultimately worth living.

An issue that continues to be a pressing one is the problem of social isolation amongst seniors. Chapter 11 returns to the participation pillar of the Active Aging Framework and explores the potential of videoconferencing to help older adults remain socially engaged. While this and other forms of ICT may be seen as potential solutions to isolation, it is important to consider the ways technologies transform social relationships, often in unintended ways. For example, many potential users of videoconferencing expressed concern about the possibility of overreliance and reduced face-to-face interaction with caregivers, family, and/or friends.

Chapter 12 provides an overview of some of the initiatives and activities in the area of gerontechnology that have emerged in Europe, North America, and the Asia-Pacific region and research at different levels: national, regional, and international. The chapter begins with a description of an international program, Everyday Technologies for Alzheimer's Care (ETAC), funded through a partnership between the Alzheimer's Association and the INTEL Corporation and designed to develop technological solutions for people with Alzheimer's disease, their caregivers, and families. The second section focuses on emerging research on technology and aging in the Asia-Pacific region. The third section focuses on national level programs and describes work being carried out in the UK funded by government agencies such as the Environment and Physical Sciences Research Council (EPSRC). The fourth section describes the very extensive program of work being funded by the European Union under the Ambient Assisted Living Joint Program. The chapter concludes by identifying some of the key common themes within the various research programs described and discusses the drivers, emerging directions, and opportunities in a rapidly expanding global research landscape.

References

Kalache, A., & Kickbusch, I. (1997). A global strategy for healthy ageing. *World Health, 4*, 4–5.
World Health Organization. (2002). *Active Aging – A policy framework*. http://whqlibdoc.who.int/hq/2002/WHO_NMH_NPH_02.8.pdf
YA2012-European Year for Active Ageing and Solidarity between Generations. http://europa.eu/ey2012/ey2012main.jsp?catId=971&langId=en

Chapter 2
Technology and the Challenge of Aging

Andrew Sixsmith

2.1   Introduction

The aging of populations is a global phenomenon. At the present time, approximately 10 % of the world's population are aged 60 years and over, but the proportion is projected to increase to about 20 % by 2050. While the population worldwide is growing at around 1 % per annum, the number of people aged over 80 is growing at 4 % per annum. Moreover, the phenomenon of population aging is not limited to the developed countries. Currently 64 % of older people live in less developed regions, but this figure will rise to 80 % by 2050 (HelpAge International, 2012).

The aging of the population presents many challenges, not least how services can be improved in order to enhance the health and quality of life of older people in an era of strained financial resources. In this context, information and communication technologies (ICTs) are viewed as having a huge potential. ICTs for older people have emerged as a major component of research and development (R&D) programs worldwide (Sixsmith, 2012). For example, the European Union's Ambient Assisted Living Joint Program has invested significantly as part of a social inclusion agenda to improve access and uptake of ICT-based products and services by disadvantaged groups, such as older people, and to exploit the opportunities this brings for European industry (EU, 2007). There is growing evidence from evaluation research that technological supports can bring about significant benefits for older people, while at the same time improving the cost-effectiveness of health and social services (Bowles & Baugh, 2007; DH, 2011; Pare, Jaana, & Sicotte, 2007). However, the research has so far been limited in terms of real-world products and services (Meyer, Muller, & Kubitschke, 2012). This chapter is an attempt to provide an agenda of research on technology and aging by firstly identifying

A. Sixsmith (✉)
Gerontology Research Centre, Simon Fraser University, Harbour Centre,
#2800-515 West Hastings Street, Vancouver, BC, Canada V6B 5K3
e-mail: Sixsmith@sfu.ca

A. Sixsmith and G. Gutman (eds.), *Technologies for Active Aging*,
International Perspectives on Aging, DOI 10.1007/978-1-4419-8348-0_2,
© Springer Science+Business Media New York 2013

opportunities for technologies for seniors and secondly by articulating some of the key challenges that will need to be addressed if the full potential of ICTs is to be exploited.

2.2 Heterogeneity and Old Age

One of the limitations of much of the research and technological development in recent years has been the way old age has been perceived and how ageist assumptions appear to have been translated into the agendas for ICTs. Old age is often seen in negative terms, equated with ill-health and disability. However, this stereotype of old age contrasts with the high degree of heterogeneity within the older population. If the potential of ICTs is to be fully exploited in terms of developing new products and services, the opportunities afforded by a growing and diverse marketplace need to be taken seriously. This section explores these opportunities in terms of four key population groups: healthy and active seniors, people with chronic diseases, people with dementia, and people with mild cognitive impairment. The focus is very much on how ICTs can be used to enhance independence and active aging and identifies some of the interesting lines of research and technological development that is beginning to emerge.

2.3 Healthy and Active Seniors

To start the discussion on the opportunities afforded by population aging for developing new technology-based products and services, it is useful to start by saying that most seniors are relatively healthy and disability-free, leading rewarding and active lives. Of course, there is a strong association between aging and health outcomes, but the majority of seniors are likely to report that they are in good health. For example, surveys in British Columbia examined functional health status in respect to vision, hearing, speech, mobility, dexterity, feelings, cognition, and pain. The majority of seniors reported *very good* or *excellent* health (Wister, Sixsmith, Adams, & Sinden, 2009), although this does vary with age, with two-thirds (66 %) of those 65–74 years reporting very good or excellent health compared with 40 % of those aged 75 years and over. Significantly, Canadian data suggest that the proportion of older people who are disabled appears to have diminished over the last few decades (Federal Interagency Forum on Aging, 2004). Data from the USA show the age-adjusted rate of disability (limitations on activities of daily living) declined from about 25 % in 1984 to 20 % in 1999; these improvements can be seen in both men and women.

Despite the relatively good health and functional status of seniors, the key focus of research and development of technologies for seniors has been on the sick and disabled. This focus appears to be limited in its vision, and there is a need to move from a technological agenda that addresses ill-health and dependency to one that promotes active aging in settings where older people want to live:

> Until recently, many traditional assumptions associated with aging and elder care often cast seniors in a more passive role, not a proactive one. For a variety of economic, sociological, and technological reasons, this paradigm is now shifting. Broadband-enabled technologies are providing seniors with an interactive lifeline to the world, empowering them to live more robust, healthful, and independent lives....With the senior population set to double in the coming decades, broadband and broadband-enabled technologies are poised to play an invaluable role in transforming senior life and the senior care paradigm. Continued competition, innovation, and investment in the broadband market will allow current and future generations of seniors to age in place, stay relevant and connected to their communities, and take advantage of lifesaving applications (Davidson & Santorelli, 2008, pp. 1–2).

The actual and potential growth of ICTs is seen by Davidson and Santorelli (2008, p. 14) to confer a number of social, economic, and health-related benefits on seniors:

- Increased connection to family and friends
- "Feelings of relevance and…an interactive outlet to the world"
- Access to e-services, such as commerce, personal finances, medication, and employment
- Improved health, wellness, and preventative care
- Enhanced health, safety, and security through telecare services
- Benefits to society at large through healthcare savings, workforce participation by seniors, and senior-related content and services

Helping older people to remain independent and *age in place* is widely recognized as important to the quality of life for individuals (Sixsmith & Sixsmith, 2008). Solutions that help older people to remain at home are of interest to health policymakers due to potential cost savings over more expensive care in institutional settings such as nursing homes (Müller & Sixsmith, 2008). In some ways, the challenge presented by healthy seniors in respect to ICTs is a social one. The agenda of dependence is strongly reflected in the research and development agendas. This has begun to shift in recent years; for example, the EU-funded Ambient Assisted Living Joint Programs (see Chap. 12) has addressed some of the more positive avenues of research and development mentioned above. A further social challenge is how the benefits of access to ICTs can be made available to all seniors. Currently, a *digital divide* in socioeconomic terms persists, where those who are mostly likely to benefit from digital products and services are the least likely to be able to access them.

2.4 People Living with Chronic Disease

Chronic diseases are physical or mental conditions that have a long-standing duration, require ongoing medical care, and usually have significant impact on the person's functional capacity and quality of life. Chronic conditions include life-threatening diseases such as heart disease, cancer, and respiratory disease.

Indeed, these have emerged as the major causes of death in the modern era as opposed to the contagious diseases of earlier periods of history. Other conditions such as arthritis and osteoporosis may not be as deadly as the major killers but are nonetheless very serious in their impact on the health and well-being of millions of seniors. For many older people, living with one or more chronic diseases is a fact of life. As the population ages, the prevalence of chronic disease also increases, presenting challenges in terms of how chronic disease is managed. Key questions to be addressed in this area are as follows: how can we help people to remain living independently at home rather than being admitted to acute care or long-term care? How can we help people to maintain their health and avoid decline into dependency? How can we help people to manage their own health conditions more effectively? How can technology help healthcare professionals monitor and communicate with patients living at home? The benefits of helping people to remain living at home are particularly apparent for this group; even when faced with declining health, older people prefer to remain living at home, while supporting people at home is a more cost-effective solution than institutional care (Wiles, Leibing, Guberman, Reeve, & Allen, 2011). Given this clear win-win scenario, then it is unsurprising that this area has received considerable attention by both academic researchers and commercial organization and a range of telehealth and telecare systems and applications have emerged that may have a very positive impact on the health and independence of this group.

One major component of emerging ICT systems is activity monitoring, which collects data from sensors in the home (e.g., infrared movement sensors) and body-worn biomedical sensors to create a profile of a person's typical pattern of living and health status, such as when they get up in the morning or how they move around the home. Research in this area has been carried out for many years, including early work by Celler, Earnshaw, Ilsar, et al. (1995), Sixsmith and Johnson (2004) and more recently Floeck, Litz, and Spellerberg (2012), and this work is now emerging as telehealth and telecare systems, although these are still a long way from being mainstream services. Activity monitoring data may be very helpful in identifying incipient health problems, facilitating early interventions and helping people to self-manage their conditions. Variations from the typical activity pattern, for example, reduced levels of activity during the day, may be indicative of a decline in health status. These systems may generate some kind of alert, for example, to a monitoring centre, which would then organize a response by healthcare professionals or informal caregivers. This approach is useful for monitoring a range of situations, for example, critical events requiring immediate response (e.g., heart attack) and non-critical situations that require longer-term preventative interventions (e.g., exacerbations of a chronic illness). Work reported by Bhachu, Hine, and Woolrych (2012) illustrates the potential of this kind of technology to help chronically ill people remain living at home. A field trial of a system using activity monitoring and biomedical sensors was set up to provide information to help community nurses to better manage patients with chronic obstructive pulmonary disease (COPD) and particularly to avoid expensive and unnecessary hospital admissions. The trial provided good evidence for the effectiveness of activity monitoring, for example, by providing easily understandable health data to the patients themselves and by prompting nurses to engage early with patients once changes in patterns of activity were detected. Early results of a large-scale trial in the UK, known as the Whole System Demonstrator, indicates that low-cost ICT-based services may have a significant benefit in terms of avoiding hospitalization reducing mortality and improving quality of life of people with long-term illnesses (DH, 2011).

Activity monitoring may also be useful in managing chronic mental health problems. For example, GPS technologies are being used to track older people with bipolar disorder in order to see whether changes in micro-geographical patterns of activity are associated with mental status (Namazi, Sixsmith, Glaesser, & O'Rourke, 2012). Bipolar disorder is characterized by rapid and extreme changes in mood, and it is hypothesized that manic and depressive periods will be reflected in the range and type of geographical activity. The system also includes the capture of *real-time in-the-moment* subjective data, where a mobile device will prompt individuals to complete a questionnaire about their mood state. Again, the use of ICT-based technologies will help care professionals to better understand the needs of patients and to provide more sensitive and timely interventions.

Despite the potential of these systems, certain concerns remain, especially in respect to the loss of privacy and the intrusiveness of being monitored at home. However, users are often willing to trade-off privacy for the additional feelings of safety and security these systems afford (Beringer, Sixsmith, Campo, Brown, & McCloskey, 2011). Older people are also often concerned that the introduction of technology will reduce the level of human interaction. It should be emphasized that these technologies cannot work in isolation and should be seen as part of an integrated care solution that enhances the formal and informal networks of care that already exist.

2.5 People with Dementia

Dementia is an age-related condition with as many as 35.6 million persons worldwide living with the condition with numbers expected to increase to 65.7 million by 2030 (Wimo & Prince, 2010). Amongst 65–70 year-olds, about 2 in every 100 will be affected, but this rises to 20 % amongst the over 80s age group. It should be emphasized that dementia is not part of the normal aging process, affecting only a fraction of the older population. However, it is a very debilitating condition for many older people leading to a severe decline in cognitive status, such as severe memory impairments, confusion, and disorientation in time and place and inability to communicate. Cognitive decline means that they can no longer do everyday activities like dressing or washing. The changes to the brain associated with conditions such as Alzheimer's disease are permanent and currently untreatable. Drugs are not yet effective in curing or significantly affecting the progress of dementia, and care interventions need to focus on managing the symptoms and providing help and support to informal carers. Typically, the application of technology for people with dementia has addressed issues of safety and security (Sixsmith, 2006).

For example, wandering is seen as a major problem for both people living in the community and in nursing homes, and there are commercially available systems to trigger an alert if a person leaves their dwelling place. Another common problem is to do with household safety, and technologies have been developed to detect and control a range of potentially dangerous situations, such as flooding, extreme temperatures, fire, and gas leaks. These are clearly important areas, but emerging AAL technologies also have the potential to develop interventions that can positively enhance independence and well-being. A major strand of research has focused on how technology can help people to perform everyday tasks of living that are important to continued independence. For example, Mihailidis, Boger, Czarnuch, Jiancaro, and Hoey (2012) have developed the COACH system using computer vision and artificial intelligence to support the user's performance of the various components of hand washing. An audiovisual interface prompts the person if help is required. Work by Kearns, Nams, and Fozard (2010) focuses on the potential of sensor technology in the assessment of cognitive status. This examines the relationship between the person's ambulatory movement and the level of cognitive function by measuring path tortuosity or the extent to which a person's movement path deviates from a straight line. Their research with residents in an assistive living facility found a significant negative association between path tortuosity and measures of cognitive status; simply put, the more erratic the pattern of movement, the lower the cognitive function of the person. This system has the potential to identify early changes in cognitive status allowing more targeted care interventions to support the person as their dementia progresses.

Other research has focused on how technology can positively enhance quality of life. For example, Sixsmith, Orpwood, and Torrington (2010) developed a music-playing device specifically for people with dementia. Music can evoke powerful memories, and people with dementia may recognize enjoy, and sing along with a song even when they may be experiencing severe problems with their memory and ability to communicate. However, access to music by people with dementia may be very limited. For example, they may not be able to operate music devices such as CD players or MP3 players. The music player was designed to be simple to use and also provided prompts to encourage people to use the device. In another project, Schmid et al. (2012) report on project MUSE, which developed a music-playing chair for use in long-term care facilities. The chair was a *smart* device that uses Radio-Frequency Identification (RFID) technology to recognize a person and play personally relevant music from a preloaded playlist. MUSE and other music-playing devices help people with dementia get access to music, and trials of the devices have been shown to have a positive impact on mood and social engagement.

2.6 People with Mild Cognitive Impairment

Declining cognitive ability has a major effect on the health and well-being of older people. As already discussed, there has been considerable attention given to the growing numbers of people with dementia. However, many older adults may be living with mild cognitive impairment (MCI), which is defined as cognitive impairment in at least one aspect of cognitive functioning, with no sign of dementia and no significant decline in functional activities of daily life (Luis, Loewenstein, Acevedo, Barker, & Duara, 2003). It has been estimated that perhaps as many as 20 % of people aged 65 and over may be experiencing MCI (Petersen, 2011). Piau, Nourhashémi, Hein, Caillaud, and Vellas (2011) point to the lack of effective pharmacological interventions for cognitively impaired people. Currently, there are no effective disease-modifying drugs, and there appear to be none in the pipeline, under trial, or near to market. This situation highlights the importance of non-pharmacological interventions to enhance the cognitive health of older people and provides a real opportunity for developing ICT-based solutions that help with planning and managing everyday activities and ensuring a safe and comfortable living environment.

ICT-related research has not yet specifically addressed the needs of people with MCI, and there has been a much greater focus on people with dementia and more severe levels of cognitive impairment. Older people with MCI have a stronger potential to learn to use new technologies than those with dementia, and introducing devices and systems for people with MCI when they are more able to adapt and interact with the technology may have more potential. Moreover, while many people with MCI will progress to dementia, this is not inevitable, and a crucial need is to develop innovative approaches that will help maintain or even improve cognitive status, extend independent living and enhance quality of life. Providing ICT *supports to enable self-care and self-management* and facilitating *cognitive stimulation* may allow those living with MCI to draw upon varying levels of support tailored to their specific needs and situations and provide stimulation and engagement to maintain or improve cognitive function. Many of the ICT solutions already described have considerable potential for people with MCI, and research funded by programs such as *Everyday Technologies for Alzheimer's Care* (see Chap. 12) has included the development of prompting technologies to remind and assist older adults in performing essential activities of daily living (ADLs). EU projects such as SOPRANO (see Chap. 8) have been developing systems to assist with tasks such as personal medication, exercise, nutrition, and home security.

One line of research which has recently emerged that may have special relevance for people with MCI is brain training, where a person engages in serious gaming, such as Nintendo's *Brain Age* to promote cognitive function. Digital games offer many potential benefits for people generally and people with MCI specifically for improving cognitive and social function in a way that is motivating and enjoyable. While the evidence base for these cognitive benefits and the understanding of the possible underlying processes remains weak, electronic games represent a huge opportunity for involving people with MCI in activities that may help them to reengage constructively with the challenges of everyday life. Astell (Chap. 10) argues that digital games can contribute to seniors' well-being through social interaction, cognitive stimulation, and physical activity that may motivate them to positively manage their lives. Games can promote highly engaged and enjoyable experiences (Csíkszentmihályi, 1996; Hwang, Hong, Hao, & Jong, 2011), offer opportunities for social interaction, and also support learning through progressive levels of practice (Sauvé & Kaufman, 2010).

2.7   The Challenge of Turning Research into Reality

Despite the considerable potential benefits of new technology outlined above, research and development in the area has had surprisingly little impact. In this section, a number of challenges to the research and development process are highlighted. While technical issues remain important, future efforts must primarily respond to a number of nontechnical challenges to service and product innovation. There is a need for a more collaborative *transdisciplinary* approach to the research and development process, while at the same time facilitating better end-user and stakeholder engagement in the research and subsequent knowledge translation. More fundamentally, there is a need for better theoretical and conceptual approaches if research is to be effective in driving technology innovation.

2.8   The Research and Development Process

It is becoming increasingly common for projects to have teams with numerous partners, involving several countries and languages, commercial, government and academic organizations and multiple disciplines. Indeed, it is difficult to envisage research in the area of technology and aging without this kind of collaboration. However, it is important to consider the many practical and operational challenges involved in the R&D process. Woolrych and Sixsmith (2012) document the problems encountered within the EU-funded SOPRANO project that aimed to develop an Ambient Assisted Living (AAL) system to help older people live independently (see Chap. 12). Multinational projects, such as SOPRANO, encounter communication barriers in terms of language, sharing information, and everyday interaction. Also, different organizations and countries present local challenges that require a level of flexibility and local autonomy that is difficult to achieve within a rigid research framework. Moreover, the different expectations of the various collaborators (including funding bodies) need to be carefully reconciled and managed throughout the R&D process. For example, how do technical problems or unexpected delays in one part of a project impact on the project as a whole, and how are they dealt with?

However, the issues of collaboration and teamwork go beyond practical and technical concerns; there is a need to look critically at the culture of research and development and the language, concepts and mindsets that frame and determine the development and implementation of new technology for older people. Research has often been driven by technological agendas and a *silo* mentality towards the development of ICT-based solutions (Müller & Sixsmith, 2008). This has been a major factor in limiting the practical outcomes, as systems and devices are often poorly aligned with the needs, preferences and wishes of end users. Inter- and multidisciplinary approaches tend to bring different disciplines together in ways which complement each other but do not generally achieve holistic understandings, focus on problem solving or explore methodological integration (Ramadier, 2004). In the area of technology and aging, where complex, multidimensional problem-based phenomena demand innovative approaches, a transdisciplinary approach is appropriate. Transdisciplinarity is characterized by a problem focus, evolving methodology and intense collaboration (Wickson, Carew, & Russell, 2006), whereby partnership working allows expertise, knowledge, and joint working practices to transcend disciplinary boundaries. Approaches such as *Action Learning* (McGill & Beaty, 2001) and *Appreciative Inquiry* (Cooperrider & Whitney, 2005) have been designed to bring together all the relevant disciplines on equal basis to engage in methodological development and interpretation of findings. Transdisciplinarity highlights the implications of research findings for technological design, conceptual development, creative processes, and professional practice much more rapidly than individual isolated efforts.

A further requirement is to educate and train a new generation of researchers and break down the traditional disciplinary boundaries (e.g., technical versus domain knowledge) and promote transdisciplinary teamworking as the basis for effective R&D. A number of these initiatives have emerged over the years. For example, the European Union funded intensive courses on gerontechnology under their Erasmus and Socrates programs. The International Society for Gerontechnology has for many years run master classes where leading researchers in the field mentor graduate students to develop their ideas, skills, and career pathways. While these initiatives are valuable, they are somewhat disparate and peripheral within the overall educational landscape, and there remains a need for a more programmatic approach, where the necessary concepts, approaches, and skills are embedded in the mainstream education of gerontologists, engineers, and computer scientists. There are signs that this is beginning to happen with various credit-bearing degree courses emerging, for example, Pace University's intergenerational service-learning computing course (http://csis.pace.edu/gerontechnology/).

2.9 User-Driven Approaches

One issue that has often been raised is that much of the research within the area of technology and aging has been technology driven, without assessing how they impact on the everyday lives of older people, often resulting in low take-up and usage of the technology by potential end users. Without a more user-focused approach, there is a danger that technologies will be ill-conceived and fail to match the needs and situations of older adults. Commentators have been largely critical of the attempts to involve users in the design and development of technology. They are rarely involved at every stage of the process, highlighting a tokenistic approach to user involvement with little representation in the development of the technology at later stages (Fudge, Wolfe, & McKevitt, 2007). Indeed, within most technology projects, the resources for user research are significantly less than those devoted to the technical aspects of the research and development. While there is still a long way to go, there has been a recent shift towards user-centered research in response to the above critique (Rubin, 1994; Steen, Kuijt-Evers, & Klok, 2007). Grounding the whole R&D process in the real experiences of users can create technologies that are usable, useful, acceptable, ethical, and supportive (Eisma, Dickinson, Goodman, Syme, Tiwari, & Newell, 2004). Rather than constraining users within rigid technological frameworks, it is important that devices are context aware and able to respond unobtrusively to the routines and activities of older people (Guo, Zhang, & Imai, 2010). The emphasis needs to be placed on putting users in control of the technology and using it to take control of their own lives and enhance choice and independence (Davidoff, Lee, Yiu, Dey, & Zimmerman, 2006).

Adopting a user-centered research approach requires a level of flexibility in the approach that can be difficult to achieve in practice. Woolrych and Sixsmith (2012), in a reflection on involving users in the development of an AAL system, highlight some of the difficulties that were encountered. In this case, the research was guided by an Experience and Application Research (E&AR) approach to involve end users (older people, carers, and service providers) in the entire R&D process (Sixsmith, 2011): requirements specification, development of use cases, prototyping, implementation, and evaluation. For example, early-stage prototypes (mock-ups) were presented by actors in a scripted drama workshop to allow an audience of older people to visulaize how the system might work and contribute their ideas to further refinement and development. Working closely with users in this way requires relationship building with local providers and older people (van den Hove, 2006), with a considerable commitment of time, effort, and resources. But beyond this, there has to be genuine participation by users and stakeholders though appropriate communication and dialogue between all the parties involved. The benefit to the research process is that users and service providers have local knowledge that is in turn helpful in recruitment and maintaining participation, for example, in field trials. Users on the other hand need to be appropriately remunerated for their involvement, as this can be a highly demanding role. There is an expectation that older users have the time and desire to participate in research, when they often lead busy lives, for example, with work, volunteering, family commitments, as well as their own personal activities. The research process can also lead to consultation fatigue that may result in them dropping out.

One of the potential benefits of a user-centered approach is that it will result in products and services that are more marketable because they accurately reflect the needs of older people. However, while technologies may be more usable and acceptable, this does not necessarily lead to innovation. A user-centered approach also involves the identification of routes to market and knowledge translation. Older users are likely to be consumers of multiple services and systems, and the development of new technologies needs to take this into account, emphasizing their complementarity and added value to existing services. It is essential to be aware of all the various stakeholders involved, such as older end users, their families, community organizations, service providers, and policymakers. These actors will often have competing agendas and needs that have to be carefully explored and accommodated within the development of technology-based solutions. For example, service-provider organizations are likely to have strongly defined cultures and processes that in themselves create challenges to the adoption of innovatory technologies. In this context, the process of technology development has to be predicated on a good understanding of the business models and processes of service providers. At this point, it is difficult to envisage that new technologies will simply replace the face-to-face care that is typical of most personal services for older people and ICTs should be seen as part of an integrated spectrum of care (Miskelly, 2001).

2.10 Knowledge Translation and Technology Exploitation

It is both interesting and ironic that despite the considerable research and development into new systems and devices to help older people, the uptake remains low despite the benefits that they can potentially provide. The *technology-push* approach has typically failed to appreciate the significant challenges to creating viable service processes and business models that include technological innovation. Making technology solutions a reality in terms of real-world products and services requires addressing these challenges in a way that creates positive outcomes for all the stakeholders involved. Indeed, our ideas of knowledge translation must go beyond the typical end-of-project dissemination approach to one that includes stakeholder participation and business modelling as fundamental to the whole R&D cycle so that technologies are congruent with the real-world opportunities and constraints. Research by Meyer and colleagues (2012) explores the international markets for the different generations of telecare systems and highlighted international differences in penetration. They argue that barriers to the creation of effective markets remain considerable worldwide, with factors such as funding and reimbursement systems and organizational issues within care services constraining service innovation. Despite this, the case for ICTs being part of the spectrum of care is increasingly compelling, but there is a need for a more holistic approach to technology development and deployment that includes key activities: user requirement analysis, care process design, and business case modelling. The aim here is to ensure that systems and devices are aligned with all the key stakeholders involved in the consumption and delivery of technology-based services.

Health and social care provision exists within a context of complex economic, social and political factors while user and financial considerations are pushing the delivery of services away from traditional settings such as long-term care facilities and hospitals into the homes and communities in which older people live. While ICTs have the potential to play an important role in the delivery of home-based services, the realization of this potential is also dependent on government policy to facilitate and encourage the uptake of new technologies within the caring services. Marin and Mulvenna (2012) examined technology policy initiatives in the UK. A key component in the effective implementation of technological solutions is a research and development cycle that is evidence-based and closely aligned to national policy objectives. In the UK, a range of initiatives have been demonstrated to have significant cost savings and improved delivery of services. The development of strong evidence-based policy has been a major driver of the introduction of new technologies in local health and social services.

The discussion in this and the previous section has focused on how to transform research and prototypes into useful real-world products and services. The key challenge is how to achieve user-centered product development alongside greater penetration of the market. This route to market is not straightforward, and indeed, some of the considerations outlined above suggest that there are contradictory factors at play. As Woolrych and Sixsmith (2012) note, providers are often highly conservative institutions that are resistive to innovation, and the ongoing failure to translate research into viable products and services points to structural weaknesses within service delivery systems as much as weaknesses in the viability of new systems and devices. ICTs offer avenues for developing innovative solutions for older people that go well beyond existing concepts of elder care, and it is beholden on those working within the field of technology and aging to adopt a critical perspective that will itself challenge the status quo rather than simply reflect the business-as-usual approach that is implicit in current lines of research.

2.11   The Need for Theory in Gerontechnology

A final issue is the way we think about technology both within the research process and in respect to how our ideas and concepts about technology and aging are reflected in the ways we support older people. In the 1980s, James Birren referred to gerontology as being data rich yet theory poor (Birren & Bengtson, 1988). While theoretical perspectives appear to have become more significant within recent gerontological literature generally, if one scans the rapidly growing literature on aging and technology, it would be fair to say that it is almost devoid of theory. There have been some attempts in the nascent field of gerontechnology to map out its disciplinary boundaries and dimensions, but even cursory comments on theoretical ideas and issues pertaining to research in the area are noticeably lacking.

Theory is important for a number of reasons. Firstly, it allows us to develop, define and articulate our basic ideas about a problem domain. Describing technology and aging research as *atheoretical* does not necessarily mean it is non-theoretical but that its theoretical ideas and principles remain largely tacit and unsystematic. Without a well-worked-out theoretical basis, there is a danger that our understanding of an area will at best remain limited and at worst embody some of the ageist attitudes that exist within society. Secondly, theories are, or should be, important drivers of empirical research, for example, in helping to develop research questions and hypotheses and to determine methods and research instruments. Theories can be used to create conceptual frameworks and models of the domain being investigated, and the development of these models is an important step to ensure that the basic approach is sound. These conceptual models are especially important when the problem is inherently complex. Thirdly, theories are useful in interpreting the results of research and are important for the advancement of knowledge in an area. Without the development of theoretical narratives (theories are essentially high-level descriptions of the how and why of a phenomena), then it is difficult to arrange our ideas and build on previous knowledge.

Without these theoretical structures, research efforts in the area of technology and aging will remain impoverished with a consequent impact in terms of technology-based products and services for older people. Potential directions for theory development include theories of innovation to explain the way new systems and devices emerge from concept to realization. A more theoretical exploration of the ideas of *needs* within the older population is also a useful avenue. For example, theoretical ideas of person–environment fit (Lawton & Nahemow, 1973) may help to align the design of technological aids to the personal and situational factors of the older user. However, the emerging field of gerontechnology remains at the very earliest stage of theoretical development. A potential starting point in the development of theoretical ideas within gerontechnology is a fourth role of theory as a means to reflect and critique on our knowledge of a problem area. The process of reflection and critique is fundamental to scientific enquiry and the advancement of knowledge and is dependent on theoretical reasoning.

This critical perspective has developed since the 1970s within gerontology (Biggs, Hendricks, & Lowenstein, 2003) with a focus on how everyday life is often structured by social processes. Within the perspective, the attitudes, roles and rules that govern everyday life are mediated language, knowledge, practices and institutions and even the physical environment in which we live—all of which are socially constructed rather than given within an immutable world. For example, the idea of *ageism* has been used to describe the way society tends to marginalize and disempower older people (e.g., through mandatory retirement), while at the same time representing this as the natural order of things. It is possible to reconceptualize technology in exactly the same way, where we can see technology not in terms of attributes such as usability or function but in social and behavioral terms, structured by socially constructed processes of development, commercialization, and patterns of usage.

At the societal level, recent years have seen significant technological change within ICT in areas such as digital media, mobile telecommunications and Internet access to information/communication, where ICTs occupy an increasingly important place in society. These technologies have significant potential benefits for socially excluded groups. Older people, for example, could benefit from access to lower-cost goods and services available online (McLean & Blackie, 2004). However, access to and engagement with ICTs is unequally distributed across society leading to a digital divide (Korupp & Szydlik, 2005). This inequality has been further analysed as digital exclusion (Cushman & Klecun, 2006), with a much larger proportion of seniors not using ICTs compared to the general population, due to factors such as lack of skills in using new technologies. Skills are acquired through both formal and informal routes to equip the older person with functional abilities in using technologies, and without them, participation in many domains of everyday life can be compromised excluding older people from wealth of potential benefits that emerge through continuing technological change.

Secondly at the behavioral level, it is important to consider the ways technologies, even those that are designed to aid old people, can in themselves contribute to social processes or marginalization. The prominent British social gerontologist, Peter Townsend, developed the concepts of structured or induced dependency to explain the way society and its institutions (e.g., retirement, poverty, institutionalization, and the restriction of social roles) contribute to the creation and maintenance of dependency within the older population (Townsend, 1986). Townsend also suggested that the dominant conceptual framework for understanding policy and practice was what he characterized as *acquiescent functionalism*:

> ... a body of thought that attributed the causes of the problems of old age to the natural consequences of physical decrescence and mental inflexibility, or to individual failures of adjustment to aging and retirement, instead of the exertions of state economic and social policy partly to serve and partly to moderate market forces. Social inequality was thereby "re-configured" in the language which is now being applied to universal social services (Townsend, 2006).

The critique was particularly aimed at a dominant social policy agenda that had emerged in the early and mid-twentieth century that had in many ways marginalized the position of older people within Western societies. A family of theoretical ideas, neoliberal economics, democratic pluralism, sociological functionalism, and ideas within social psychology implicitly saw the stages of industrial development and modernization as progressive, benign, and beneficial for society. The acceptance of these ideas reinforced a concept of old age that accepted the structural inequalities within society and the dependency of older people as an inevitable consequence of this process and the combined ideas within acquiescent functionalism served to validate a policy of dependency. While a very strong critique of this perspective emerged in the later decades of the twentieth century, there remain features of institutionalized agism that persist in shaping the lives of older people in contemporary society.

As mentioned, the underlying idea from critical gerontology is to try and reconceptualize technology as something that is fundamentally social in nature. For example, much of the technology development effort has been focused on enhancing the care and support provided to older people. Sixsmith (2012) suggests that the increasing use of surveillance technologies to monitor the activities and health status of older people is justified on the grounds of providing help and support to those in need. However, little attention has been given to how these technologies might impact on users' everyday lives beyond technical issues of confidentiality and data protection. While these systems ostensibly aim to help older adults live more safely, their development and deployment has generally adopted an *agenda of dependence*, focusing on issues of clinical needs, impairment, and incapacity. This is particularly important given that monitoring technologies have extended surveillance into the fabric of everyday life. Even though users are generally accepting of these technologies, the intrusion of the *clinical gaze* into everyday life changes the power relationships between the carer and the client and observer and observed. The exposure of hidden, intimate spaces to external observation has a regulatory effect (Foucault, 1995) both on the everyday behavior of the person being observed and the actions by the observer/carer necessitated by the new categories of knowledge afforded by surveillance.

Technology-based solutions tend to be constrained within an existing provider-led agenda of care that has typically focused on reactive services rather than preventative interventions, and an uncritical adoption of technology may actually contribute to a process of induced dependency. Moreover, the technology/care agenda also fails to recognize the way technologies can transform everyday life and social relationships. Personal and social constructs and characteristics such as gender, education, ethnicity, and disability play key roles in older adults' relationship with technology, impacting on their social inclusion and quality of life. A vital element in understanding this is the older person's own views and experiences of technology, which can often be at odds with mainstream conceptualizations (Freeman & Lessiter, 2001; Sixsmith & Sixsmith, 2000). In particular, technology is often framed in terms of *supporting* frail and disabled older people. While this ostensibly aims to help older adults live independently and enhance their quality of life, the approach has generally focused on issues of *impairment*, *age-related decline*, and the *problems* associated with later life rather than positive aging. Such research largely fails to address the very diverse experiences of older people and their meaningmaking around technology in everyday life, although exceptions to this model of technology and decline do exist, for example, McCreadie and Tinker (2005) identify older adults' own meanings of the consequences of *assistive* technologies for their identity. This work suggests that people's life experiences and *technological biographies* within the context of the lifecourse may also impact on their engagement with a changing technological landscape (Avgerou & McGrath, 2007). Further, much of the everyday experience of interacting with technologies happens at the level of practical activity embedded amongst all other everyday experience. The key issue here is identifying *how* people engage with an everyday world that is profoundly familiar and largely taken for granted, so that the findings can become the basis for developing suitable technological interventions that can be embraced by older adults because they can readily recognize their potential for maintaining a good quality of life and also feel well equipped to utilize them in the ways, and for the purposes, that suit the individual within the context within which they live.

2.12 Conclusion

The fact that most people in the developed world (and increasingly in the developing world) can now expect to live into old age is one of the great success stories of recent history. In many ways, modern society and its political and economic institutions are still coming to terms with the very dramatic progress that has been made in the last 100 years or so in terms of reduced mortality and fertility rates, better standards of living, health, and hygiene, and rapid advances in medical treatment. However, the discourse on aging and old age within society is often highly negative. While older people often experience a range of problems associated with the aging process, their depiction is typically stereotyped and overgeneralized and fails to recognize the positive experiences of older individuals and their contributions to society.

The technological solutions outlined in the first part of this chapter highlight the many potential ways information and communication technologies can be harnessed to enhance the well-being of older adults. However, the research agendas that have emerged may in turn be constrained by a number of practical, methodological, and conceptual weaknesses. Perhaps the most significant is that technology research has often used the same language of problematization and desperation that has been described as an *apocalyptic gerontology* that is informed by a discourse of dependency and the overwhelming impact of aging on modern societies (Gee & Gutman, 2000). The idea of *dependency* characterizes old age in terms of illness, frailty, impairment, and incapacity, implying a simple relationship between population aging and the demands placed on health and welfare services. However, as Lloyd-Sherlock, McKee, Ebrahim, Gorman, Greengross, Prince, et al. (2012) highlight, simple measures of health (e.g., morbidity) are not always good indicators of functional status. The relationship between health, functional status (the ability to perform tasks of daily living), and the demands placed on healthcare systems is far from straightforward. For instance, there are considerable variations within and between countries in the health and functional status of older populations. Lloyd-Sherlock and colleagues also point to the mounting evidence that relatively cheap and simple interventions can have a very significant impact on the person's ability to live independently and manage their health conditions at home and better information and advice about health issues, preventative measures, and support for carers can all have a major impact. However, most countries have failed to make these kinds of interventions generally available, representing almost a negligent attitude within the policy and practice arenas to make an impact on the health and well-being of many millions of people, while at the same time creating much of the service demand for which the rising population of older people is blamed (Lloyd-Sherlock et al., 2012):

> …instead of being portrayed as a problem, increased human longevity should be a cause for celebration. Moreover, population aging provides opportunities to rethink health policy for the benefit of all (Lloyd-Sherlock et al., 2012).

Much of the discussion in this chapter has aimed to defuse the so-called demographic timebomb and to emphasize how we can potentially utilize technology for the betterment of society by enhancing the independence, health and quality of life of older people. Yet it is important not to underestimate the huge shift this will require in the mindset about old age and the reorienting of the research and development agenda. While technical challenges still exist, the most significant and intransigent challenges are likely to be more social and attitudinal in nature.

References

Avgerou, C., & McGrath, K. (2007). Power, rationality, and the art of living through socio-technical change. *MIS Quarterly, 31*(2), 295–315.

Beringer, R., Sixsmith, A., Campo, M., Brown, J., & McCloskey, R. (2011). The "acceptance" of ambient assisted living: Developing an alternative methodology to this limited research lens. In *Lecture Notes in Computer Science: Proceedings from the 9th International Conference on Smart Homes and Health Telematics LCNS 6719–ICOST 2011*. Berlin: Springer.

Bhachu, A., Hine, N., & Woolrych, R. (2012). The role of assistive technology in supporting formal carers. In J. Augusto et al. (Eds.), *Handbook of ambient assisted living* (pp. 283–303). Amsterdam: IOS Press.

Biggs, S., Hendricks, J., & Lowenstein, A. (2003). The need for theory in gerontology. Introduction. In S. Biggs, A. Lowenstein, & J. Hendricks (Eds.), *The need for theory: Critical approaches to social gerontology* (pp. 1–12). New York: Baywood.

Birren, J. E., & Bengtson, V. L. (1988). *International perspectives on families, aging and social support*. New York: Springer.

Bowles, K. H., & Baugh, A. C. (2007). Applying research evidence to optimize telehomecare. *Journal of Cardiovascular Nursing, 22*, 5–15.

Celler, B. G., Earnshaw, W., Ilsar, E. D., Betbedermatibet, L., Harris, M. F., Clark, R., et al. (1995). Remote monitoring of health status of the elderly at home. A multidisciplinary project on aging at the University of New South Wales. *International Journal of Bio-Medical Computing, 40*(2), 147–155.

Cooperrider, D., & Whitney, D. (2005). *Appreciative inquiry: A positive revolution in change*. San Francisco, CA: Case Western Berrett-Koehler.

Csíkszentmihályi, M. (1996). *Creativity: Flow and the psychology of discovery and invention*. New York: Harper Perennia.

Cushman, M., & Klecun, E. (2006). *Non-users of computers in South London. Their experiences and aspirations for use*. London: Department of Information Systems, London School of Economics and Political Science.

Davidoff, S., Lee, K., Yiu, C., Dey, A., & Zimmerman, J. (2006). Principles of smart home control. In P. Dourish & A. Friday (Eds.), *Ubicomp* (Lecture Notes in Computer Sciences, Vol. 4206, pp. 19–34). Berlin: Springer.

Davidson, M. C., & Santorelli, J. M. (2008). *Network effects: An introduction to broadband technology & regulation*. Washington, DC: U.S. Chamber of Commerce.

DH. (2011). *Whole Systems Demonstrator Project: Headline figures*. http://www.dh.gov.uk/en/Publicationsandstatistics/Publications/PublicationsPolicyAndGuidance/DH_131684

Eisma, R., Dickinson, A., Goodman, J., Syme, A., Tiwari, L., & Newell, A. F. (2004). Early user involvement in the development of Information Technology-related products for older people. *Universal Access in the Information Society, 3*(2), 131–140.

EU. (2007). *European i2010 initiative on e-Inclusion*. Retrieved January 6, 2012 from http://ec.europa.eu/information_society/activities/einclusion/docs/i2010_initiative/comm_native_com_2007_0694_f_en_acte.pdf

Federal Interagency Forum on Aging. (2004). *Older Americans 2004*. Washington, DC: US Government Printing Office.

Floeck, M., Litz, L., & Spellerberg, A. (2012). Monitoring patterns of inactivity in the home with domotics networks. In *Handbook of ambient assisted living* (pp. 258–282). Amsterdam: IOS Press.

Foucault, M. (1995). *Discipline and punish* (2nd ed.). New York: Vintage Books.

Freeman, J., & Lessiter, J. (2001). ITC-UsE Ease of Use and Knowledge of Digital and Interactive Television: Results. Independent Television Commission: http://www.ofcom.org.uk/static/archive/itc/uploads/UsE_report.pdf.

Fudge, N., Wolfe, C. D. A., & McKevitt, C. (2007). Involving older people in health research. *Age and Ageing, 36*, 492–500.

Gee, E. M., & Gutman, G. M. (2000). *The overselling of population aging: Apocalyptic demography, intergenerational challenges, and social policy*. Don Mills, ON: Oxford University Press.

Guo, B., Zhang, D., & Imai, M. (2010). Enabling user-oriented management for ubiquitous computing. *Journal of Computer Networks, 54*(16), 2840–2855.

HelpAge International. (2012). http://www.helpage.org/resources/ageing-data/

Hwang, M.-Y., Hong, J.-C., Hao, Y.-W., & Jong, J.-T. (2011). Elders' usability, dependability, and flow experiences on embodied interactive video games. *Educational Gerontology, 37*(8), 715–731.

Kearns, W., Nams, V., & Fozard, J. (2010). Wireless fractal estimation of tortuosity in movement paths related to cognitive impairment in assisted living facility residents. *Methods of Information in Medicine, 49*(6), 592–598.

Korupp, S. E., & Szydlik, M. (2005). Causes and trends of the digital divide. *European Sociological Review, 21*(4), 409–422.

Lawton, M. P., & Nahemow, L. (1973). Ecology and the aging process. In C. Eisdorfer & M. P. Lawton (Eds.), *Psychology of adult development and aging* (pp. 657–668). Washington, DC: American Psychological Association.

Lloyd-Sherlock, P., McKee, M., Ebrahim, S., Gorman, M., Greengross, S., Prince, M., et al. (2012). Population ageing and health. *The Lancet, 379*(9823), 1295–1296. doi:10.1016/S0140-6736(12)60519-4.

Luis, C. A., Loewenstein, D. A., Acevedo, A., Barker, W. W., & Duara, R. (2003). Mild cognitive impairment: directions for future research. *Neurology, 61*, 438–444.

Marin, S., & Mulvenna, M. (2012). Delivering technology-enriched health and social care: Policy context for user-focused research. In J. Augusto et al. (Eds.), *Handbook on AAL for healthcare, well-being and rehabilitation* (pp. 369–386). Amsterdam: IOS Press.

McCreadie, C., & Tinker, A. (2005). The acceptability of assistive technology to older people. *Ageing and Society, 25*, 91–110.

McGill, I., & Beaty, L. (2001). *Action learning: A guide for professional management and educational development*. London: Routledge.

McLean, R., & Blackie, N. M. (2004). Customer and company voices in e-Commerce: A qualitative analysis. *Qualitative Market Research: An International Journal, 7*(24), 243–249.

Meyer, I., Muller, S., & Kubitschke, L. (2012). AAL markets: Knowing them, researching them. Evidence from European Research. In J. Augusto et al. (Eds.), *Handbook on AAL for healthcare, well-being and rehabilitation* (pp. 346–368). Amsterdam: IOS Press.

Mihailidis, A., Boger, J., Czarnuch, S., Jiancaro, T., & Hoey, J. (2012). Ambient assisted living technology to support older adults with dementia with activities of daily living. In J. Augusto et al. (Eds.), *Handbook of ambient assisted living* (pp. 304–330). Amsterdam: IOS Press.

Miskelly, F. (2001). Assistive technology in elderly care. *Age and Ageing, 30*, 455–458.

Müller, S., & Sixsmith, A. (2008). User requirements for ambient assisted living: Results of the SOPRANO project. *Gerontechnology, 7*(2), 168.

Namazi, S., Sixsmith, A., Glaesser, U., & O'Rourke, N. (2012, June). *Developing a mobile experience sampling tool for seniors with bipolar disorder.* Paper presented at the World Conference of the International Society of Gerontechnology, Eindhoven.

Pare, G., Jaana, M., & Sicotte, C. (2007). Systematic review of home telemonitoring for chronic diseases: The evidence base. *Journal of the American Medical Informatics Association, 14*, 269–277.

Petersen, R. C. (2011). Mild cognitive impairment. *New England Journal of Medicine, 364*, 2227–2234.

Piau, A., Nourhashémi, F., Hein, C., Caillaud, C., & Vellas, B. (2011). Progress in the development of new drugs in Alzheimer's disease. *Journal of Nutrition, Health and Aging, 15*(1), 45–57.

Ramadier, T. (2004). Transdisciplinarity and its challenges: The case of urban studies. *Futures, 36*(4), 423–439.

Rubin, J. (1994). *Handbook of usability testing*. New York: Wiley.

Sauvé, L., & Kaufman, D. (Eds.) (2010). *Jeux et simulations éducatifs: Etudes de cas et leçons apprises* [Educational games and simulations: Case studies and lessons learned]. Saint-Foi, Québec: Presses de l'Université du Québec.

Schmid, I., Marks, B., Sixsmith, A., Jung, K., Baines, S., & Carson, J. (2012). Music player for dementia care homes. *Gerontechnology, 11*(2), 419.

Sixsmith, A. (2006). New technologies to support independent living and quality of life for people with dementia. *Alzheimer's Care Quarterly, 7*(3), 194–202.

Sixsmith, A. (2011). A user-driven approach to developing ambient assisted living systems for older people. In J. Soar, R. Swindell, & P. Tsang (Eds.), *Intelligent technologies for bridging the grey divide* (pp. 30–45). New York: Information Science.

Sixsmith, A. (2012). International initiatives on development of information and communication technologies to assist older people. *Gerontechnology, 11*(2).

Sixsmith, A., & Johnson, N. (2004). A smart sensor to detect falls of the elderly. *IEEE Pervasive Computing, 3*(2), 42–47.

Sixsmith, A., Orpwood, R., & Torrington, J. (2010). Developing a music player for people with dementia. *Gerontechnology, 9*(3), 421–427.

Sixsmith, A., & Sixsmith, J. (2000). Smart home technologies: meeting whose needs? *Journal of Telemedicine and Telecare, 6*(1) Supplement, 190–192.

Sixsmith, A., & Sixsmith, J. (2008). Ageing in place in the United Kingdom. *Ageing International, 32*(3), 219–235.

Steen, M., Kuijt-Evers, L., & Klok, J. (2007). *Early user involvement in research and design projects – a review of methods and practices*. Paper for the 23rd EGOS Colloquium (European Group for Organizational Studies).

Townsend, P. (1986). Ageism and social policy. In C. Phillipson & A. Walker (Eds.), *Ageing and social policy: A critical assessment* (pp. 15–44). Aldershot, Hampshire: Gower.

Townsend, P. (2006). Policies for the aged in the 21st century: More 'structured dependency' or the realisation of human rights? *Ageing and Society, 26*, 161–179.

van den Hove, S. (2006). Between consensus and compromise: acknowledging the negotiation dimension in participatory approaches. *Land Use Policy, 23*, 10–17.

Wickson, F., Carew, A. L., & Russell, A. W. (2006). Transdisciplinary research: Characteristics, quandaries and quality. *Futures, 38*(9), 1046–1059.

Wiles, J. L., Leibing, A., Guberman, N., Reeve, J., & Allen, R. E. S. (2011). The meaning of "aging in place" to older people. *The Gerontologist*. doi:10.1093/geront/gnr098.

Wimo, A., & Prince, M., (2010). The global economic impact of dementia. *ADI World Alzheimer Report 2010*.

Wister, A., Sixsmith, A., Adams, R., & Sinden, D. (2009). *Fact book on aging in British Columbia* (5th ed.). Vancouver, BC: Gerontology Research Centre, Simon Fraser University.

Woolrych, R., & Sixsmith, A. (2012). Challenges of user-centred research in the development of ambient assisted living systems. In M. Donnelly, C. Paggetti, C. Nugent, & M. Mokhtari (Eds.), *Impact analysis of solutions for chronic disease prevention and management* (Lecture Notes in Computer Science, Vol. 7251, pp. 1–8). Berlin: Springer.

Chapter 3
Older Adults and the Adoption of Healthcare Technology: Opportunities and Challenges

Sara Czaja, Scott Beach, Neil Charness, and Richard Schulz

3.1 Introduction

The population is aging at an unprecedented rate in both developed and developing countries. The number of people aged 65 and over worldwide was 506 million in 2008 and is projected to be around 1.3 million by 2040 (Kinsella & He, 2009). In the USA alone by 2030 there will be about 72 million people aged 65 and over who will represent 19.3% of the population. A critically important feature of population aging is the growth in the number of people aged 80 and older who represent the oldest old (Fig. 3.1). Worldwide the 80+ population is expected to increase by 233%. In the USA people age 80+ will number about 6.6 million by 2020 and will represent 35% of the older population by 2040.

The growth in the number of older people especially those who represent the *oldest old* has vast implications for the healthcare system as the likelihood of developing a chronic disease or disability, and the need for healthcare services generally

S. Czaja, Ph.D. (✉)
University of Miami Miller School of Medicine, Miami, FL, USA
e-mail: SCzaja@med.miami.edu

S. Beach, Ph.D. • R. Schulz, Ph.D.
University Center for Social and Urban Research, University of Pittsburgh,
121 University Place, Pittsburg, PA 15260, USA
e-mail: scottb@pitt.edu; schulz@pitt.edu

N. Charness, Ph.D.
William G. Chase Professor of Psychology, Florida State University,
1107 West Call Street, Room A205, Tallahassee, FL 32306-4301, USA
e-mail: Neil_charness@embarqmail.com

A. Sixsmith and G. Gutman (eds.), *Technologies for Active Aging*,
International Perspectives on Aging, DOI 10.1007/978-1-4419-8348-0_3,
© Springer Science+Business Media New York 2013

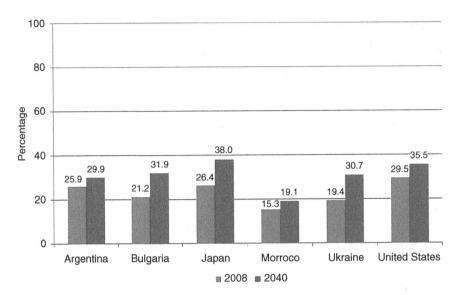

Fig. 3.1 Oldest old as a percentage of all older people: 2008 and 2040 (*Source*: An Aging World: 2008. U.S. Census Bureau, International Population Reports, P95/09-1. U.S. Government Printing Office, Washington, DC (2009))

increases with age. For example, within the USA, about 80% of older adults have a chronic condition such as heart disease, diabetes, or arthritis, about 50% have at least two conditions (Center for Disease Control and Prevention, 2010), and large numbers of older people have functional limitations that interfere with the performance of daily living tasks (Fig. 3.2). Further, about 5 million Americans aged 65 and older have Alzheimer's disease, and this number will increase in the coming years especially with the growth in the oldest old (Alzheimer's Association, 2010).

Clearly strategies are needed to ensure that older adults and families who are in need receive the care and services they require and to promote health, well-being, and independence among older people. Poor health is not an inevitable consequence of growing older. Although there is a greater propensity towards developing chronic conditions and disabilities with age, current generations of older adults are in many ways healthier than prior generations. In addition, many older adults lead active lives and are actively involved in the management of their health. As discussed below healthcare technologies provide opportunities for enhancing the ability of older people to be actively engaged in health self-management.

At the same time that the population is aging, there are marked changes occurring within healthcare systems. In this new environment, individuals and their families are expected to assume an increasing role in the management of their own health, perform a range of healthcare tasks, and interact with a vast array of medical devices and technologies within home and community settings. For example, electronic links between healthcare professionals and older patients provide healthcare providers with easier access to their patients and allow them to conduct daily status

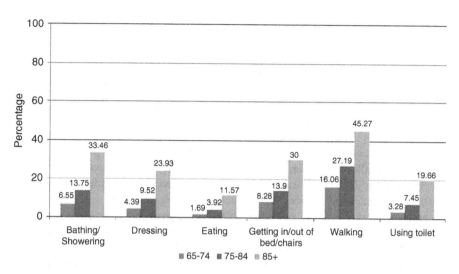

Fig. 3.2 Percentage of persons with disabilities in activities of daily living by age group: 2007 (*Source*: A Profile of Older Americans: 2010. Administration on Aging. U.S. Department of Health and Human Services)

checks or to remind patients of home healthcare regimes. Technology applications are also commonly used within home settings to monitor a patient's physical, emotional, or cognitive functioning. Current technologies can also facilitate the ability of caregivers to monitor older relatives who are in need of care and support. These applications offer the potential of allowing many people who are at risk for institutionalization to remain at home. In addition, with the rapid introduction of electronic medical records (EMRs), many of which have patient portals, patients will have access to varying degrees of their medical information, perform tasks such as communicate electronically with providers, schedule appointments, renew prescriptions, and access health management information through links to medical websites. There are also a myriad of health websites available that provide consumers with access to health information and services and the ability to buy medical supplies, equipment, and even medications/supplements. Technology also offers opportunities for increased social connectivity which can be extremely beneficial for many older people such as those who are isolated or live alone or geographically distant from family members. Social isolation is associated with poorer quality of life, life satisfaction, and well-being, poorer health status, and distress and mental illness (Cantor & Sanderson, 1999; Cobb, 1976; Dykstra, 1995; Ellaway, Wood, & MacIntyire, 1999; Ellis & Hickie, 2001).

Overall, technology holds the promise of improving access to health care for older people and empowering them to take an active role in health self-management. However, although the rate of technology adoption is increasing among older people, existing data indicate that there are still age-related gaps in usage. For example, in the USA in 2010 about 42% of people age 65+ were Internet users as compared to 78% of people age 50–64 and 87% of those 30–49 years old (U.S. Census Bureau,

2011). Older adults using computers and Internet tend to be better educated, white, have greater social resources, and fewer functional impairments than non-adopters. Home broadband adoption is also lower among older people. In 2010, only 31% of people aged 65+ had broadband access at home which limits the scope and potential of the online experience and the ability to use many health applications (Smith, 2010). Use of technology also tends to be lower among people with chronic conditions. According to recent data from the Pew Internet & American Life Project (Fox & Purcell, 2010), living with a chronic disease has an independent negative effect on someone's likelihood to have Internet access especially if they have more than one chronic condition. Further, although 82% of adults in the United States have a cell phone, only 57% of people aged 65+ years report cell phone ownership (Lenhart, 2010). Communication devices are increasingly becoming more integrated with computer network resources providing faster more powerful interactive services, and an increasing number of health services are becoming available for mobile devices. The wide deployment of technology within health care implies that lack of technology use will have increasingly negative implications and contribute to healthcare disparities. For the full benefits of health technologies to be realized for older people and their families, it is necessary to understand user characteristics, needs, and preferences in order to maximize the usefulness and usability of these technologies for these populations. This chapter will discuss the potential of healthcare technologies for older adults, factors that influence the adoption and use of these technologies, and strategies to enhance technology uptake. An emphasis will be given to health applications on the Internet and monitoring technologies given the broad use of these applications and their potential for improving the health and independence of older adults.

3.2   E-Health Applications and Older Adults

In the last decade the rapid growth in communication technologies and the Internet has created new possibilities for individuals to assume a more pronounced role in their own health and health care. In fact, recently the term *e-health* has emerged which refers to the interaction of an individual (e.g., consumer, patient, healthcare professional) with digital information and communications technologies such as the Internet and mobile devices to access or receive health information, guidance, or support on a health-related issue (US Department of Health and Human Services, 2011a, 2011b). There are numerous tools and resources that fall under e-health including social networks and support groups; online health information; online health self-management tools; and online access to personal health records.

One of the most common forms of e-health is health information seeking by consumers. A number of websites are available that provide consumers with information on illnesses/diseases, medications and treatments, healthcare providers, and health resources. Social networks and blogs related to health issues are also becoming popular. Government agencies are also increasingly using the Internet to exchange information and for services. In the USA, 80% of Internet users and 57% of the total adult population search for health information online from health websites. The most common types of information searched for are information about a disease or medical problem, medical treatment or procedure, or information about doctors or other health professionals (Fox, 2011a). Seeking health advice or receiving support from peers online is also becoming a significant source of health information.

In general, although people with chronic conditions are less likely to go online than those without chronic conditions, those with chronic conditions that are online are avid seekers of online health information. Twenty-three percent of Internet users living with a chronic condition such as high blood pressure, diabetes, heart disease, or cancer indicate that they have gone online to find others with similar health concerns as compared to 15% of Internet users without a chronic condition. However, overall older adults and those with lower educational and income levels are less likely to look for health information online than are other groups (Fox, 2011b).

The Internet can be very beneficial to older adults as it provides them with the opportunity to become more informed and thus better prepared to discuss treatment plans with their physicians (Taha, Czaja, & Sharit, 2009), seek advice from others, for example, from other medical experts or organizations or through social networking sites directed at people with similar health issues, and explore the risks and benefits associated with decision options. The Internet may also be beneficial for family caregivers who are providing care for an older person with a chronic illness or disease such as dementia. Networks can link caregivers to each other, healthcare professionals, community service, and educational programs. Recent findings from an interview study of approximately 1,500 caregivers indicated that 53% use the Internet as a source of information about caregiving (National Alliance for Caregiving, 2009). The following section discusses some of the potential issues associated with health information seeking for older adults.

3.3   Issues Surrounding Online Health Information Seeking and Older Adults

Although the Internet holds promise in terms of improving access to healthcare and health information for older adults, to date many Internet-based health applications have been designed without consideration for needs, capabilities, and preferences of this user group. For example, the US government's Medicare.gov website is intended to support health-related activities, such as finding information and solving problems related to healthcare benefits. However, a recent investigation of the usability of this website revealed that, despite having had experience searching for information on the Internet, older study participants encountered greater difficulty and generally performed poorly using it, compared to younger users (Czaja, Sharit, & Nair, 2008). We also conducted focus groups to gain insight regarding the health information needs of older adults and sources of health information and to determine if there are differences in perceptions and use of health information between Internet and non-Internet users (Taha et al., 2009). Overall, the Internet users had very positive perceptions about health information online and indicated that access to the information increased their ability to take care of themselves. For those participants who did not use the Internet for health information, the most common reasons for nonuse were related to lack of skill, concerns about security and the quality of the information, and perceptions that the Internet is too complicated.

One major concern within the e-health arena is the vast amount of information that is available to consumers. For example, in September 2011 if one typed the word dementia into the Google search box, there were 40,200,000 hits and arthritis resulted in 88,400,000 hits. This can be daunting for older adults, many of whom have limited Internet experience and limited knowledge of credible sources of information. The credibility of information on health websites varies considerably. Results from the Pew Internet & American Life Project (2008) indicate that most consumers do not consistently check the source and date of health information they find online. Other concerns related to the ability of consumers to integrate and interpret the wealth of available information. The content on many health-related websites is highly technical and difficult for nonmedical specialists to understand. Our data (Czaja et al., 2006) indicate that older adults often have difficulty interpreting information provided on health websites and often find the language to be difficult to understand. Information seeking is an activity that places demands on many cognitive abilities. One of our studies (Sharit, Hernandez, Czaja, & Pirolli, 2008) found that reasoning, working memory, and perceptual speed were significant predictors of performance of health-related Internet search. This is potentially problematic for older adults as these abilities tend to decline with age. Our data indicated that older adults who had better performance had higher cognitive abilities than those who performed at a lower level (Czaja, Sharit, Hernandez, Nair, & Loewenstein, 2010).

Clearly interventions are needed to enhance the accessibility and usability of Internet-based health applications for older adults. There are guidelines available that can be directed at making websites usable for older adults (Zaphiris, Ghiawadwala, & Mughal, 2005). For designers, however, adherence to these guidelines is not always straightforward, especially with guidelines that are general in nature and deal with issues related to the cognitive demands associated with websites. Clearly more research is needed to identify strategies and tools to help people effectively filter, organize, and integrate information. In addition, it would be helpful if designers had basic information on the needs and preferences of older adults and age-related changes in abilities that have relevance to the design of healthcare technologies. In this regard, designers should also adopt a user-centered approach to design and include representative samples of older adults in product usability evaluations.

3.4   Monitoring Technologies

3.4.1   The Potential of Monitoring Technology for Older Adults

Technologies that monitor behavior and communicate with professionals and family members offer great promise for enabling older adults to maintain independence by *aging in place* and to ultimately enhance quality of life. Such systems could, for example, know how well a person slept last night, identify potential health problems before they become serious or catastrophic, know whether they are able to carry out daily routines, and assure a daughter who lives in a distant city that they are doing well today. Various monitoring systems for older adults and their caregivers are already on the market, and many more are being developed.

Home-based monitoring technology for older adults offers a variety of potential benefits. The Center for Aging Services Technologies (CAST) categorizes these systems into three broad domains (1) safety, (2) health and wellness, and (3) social connectedness (Alwan, Wiley, & Nobel, 2007). Home-based safety monitoring technologies include *fall detection and prevention* systems, both push-button and accelerometer-based wearable (e.g., Life Alert) and sensor-based embedded environmental systems (e.g., QuietCare); *mobility aids* for wheelchairs (e.g., to enable stair climbing) and robotic walkers to enhance safe navigation; *stove use detectors* (e.g., Stove Guard); and *smoke and temperature monitors*. Health and wellness monitoring technologies include *wearable activity monitors* using accelerometers and sensors (e.g., Bodymedia); *non-wearable, embedded sensor activity monitors* to track activities of daily living (ADL), instrumental activities of daily living (IADL), and other behaviors (e.g., Healthsense); *hybrid wearable/environmental* systems with radio-frequency identification (RFID) readers and tagging of environmental objects to monitor ADL performance; *ambulatory monitors* to record and transmit physiological data (e.g., cardiac event and Holter monitors); *passive environmental non-wearable* systems like bed monitors for clinical sleep assessment; *medication compliance* systems that monitor intake and provide prompts and reminders; and *cognitive assessment/orthotics* devices.

Potential benefits of health and wellness monitoring for older adults include improved health outcomes and quality of life, empowerment and self-directed health, and prolonged independence, while informal caregivers may benefit from being more informed about their loved one's health, improved health-related communication, opportunities for prevention and early detection and intervention, and reduced burdens and strains of care. It should be noted that while these systems are becoming more commonplace, evidence for their actual impact on caregivers and older adults' health outcomes is generally weak (Alwan et al., 2007). Social connectedness monitoring is a relatively new area of application and involves the use of sensors to facilitate awareness and interaction between remote family members (e.g., INTEL's presence lamp). Older adults may benefit from improved quality of life via increased social interaction, and reduced isolation, with the potential for improved health outcomes as a result, while informal caregivers can benefit from improved communication with their loved one.

A more recent development is the design and implementation of *smart home* applications. These involve integrated networks of sensors—which may include a combination of safety, health and wellness, and social connectedness technologies—installed into homes or apartments to simultaneously and continuously monitor environmental conditions, daily activity patterns, vital signs, sleep patterns, etc. over the long term. One example is the system being developed at the University of Missouri (*TigerPlace*) by Skubic, Alexander, Popescu, Rantz, and Keller (2009). The goal is to capture physical and cognitive behavioral patterns and develop algorithms to detect deviations from normal patterns in the hopes of early detection of health problems and prevention of health declines. The potential benefits of such technology are evident, although some would argue that the constant monitoring and "big brother" qualities of whole-home systems may outweigh any benefit and lead to low acceptance or abandonment of the technology. The next section discusses more general issues surrounding older adults' potential acceptance of monitoring technology.

3.4.2   Issues Surrounding Acceptance of Monitoring Systems by Older Adults

All technology, including the monitoring technologies discussed here, involves potential barriers to acceptance that must be overcome to facilitate widespread acceptance, adoption, and continued use. These include a broad range of user characteristics (socio-demographics, health status, social support, experience with and attitudes towards technology) and resources (sensory, cognitive, psychomotor); system characteristics (user interface, instructional support, aesthetics, engagement, functionality); and the fit between the user and the system. This section discusses a reduced set of the key potential barriers to acceptance of monitoring technologies by older adults. We start the discussion with the issue of privacy, which is certainly a potential stumbling block for technology that involves monitoring, surveillance, and potential sharing of data on a range of behaviors and vital signs with family, healthcare providers, and others. We present some of our own work in this area with surveys of disabled and nondisabled baby boomers and older adults, followed by discussion of a recently developed privacy framework that confirms and extends our work (Lorenzen-Huber, Boutain, Camp, Shankar, & Connelly, 2010). The section concludes with a discussion of the other key issues for the acceptance and adoption of monitoring technologies by older adults. These include perceived need for the technology by the older adult, perceived system usefulness, system demands, and the effects of monitoring technologies on social interaction and social connectedness.

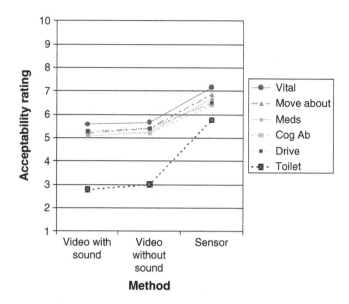

Fig. 3.3 Acceptability of recording information from different domains using varying methods

Privacy: Monitoring technologies can potentially create concerns about *informational privacy*—*what* type of information is recorded, *how* it is recorded, and with *whom* it is shared. In a recent national Web-based survey of 1,518 disabled and nondisabled baby boomers (age 45–64) and older adults (age 65+), our research group found variation in attitudes across these dimensions (Beach, Schulz, Downs, Matthews, Barron, et al., 2009; Beach et al. 2010). We varied what type of information was being recorded (vital signs, moving about the home, taking medications, cognitive abilities, driving behavior, toileting) and crossed this with three methods of recording (video with sound, video without sound, sensors) separately with the target recipient of the information (self, family, doctor, researchers, insurance companies, government). Figure 3.3 shows potential users were less accepting of the use of video cameras, either with or without sound, than of sensors (using a 10-point scale with 1=completely unacceptable and 10=completely acceptable). Note that the acceptability of recording toileting behavior via video was very low, although recording this highly sensitive activity with sensors was seen as moderately acceptable.

Figure 3.4 shows less acceptance of sharing information about toileting behavior and, to a lesser extent, driving behavior and that insurance companies and the government are least acceptable as potential recipients of health information, while family members and doctors are most acceptable. Also, note that sharing information about driving behavior is less acceptable outside of family contexts. The other major finding of the study was that both baby boomers and older adults reporting higher levels of disability were more accepting of having information recorded and

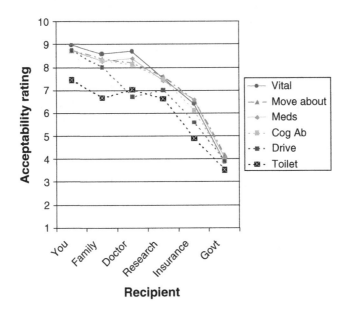

Fig. 3.4 Acceptability of sharing information from different domains with varying recipients

shared than those with lower levels or no disability. We found a dose–response effect where those reporting both ADL and IADL difficulties were most accepting, followed by those reporting only IADL difficulties, and followed by those who were nondisabled. The study provided empirical evidence of the implicit trade-offs between reduced privacy and improved health and suggested that such trade-offs may be more likely among those most in need of help (Beach et al., 2010; Beach, Schulz, Downs, Matthews, Barron, et al., 2009).

In a follow-up survey of 403 gerontology research registry and 217 wheelchair registry members aged 45 and older, our group explored privacy trade-offs and monitoring technology more explicitly and extensively (Beach, Schulz, Downs, Matthews, Seelman, et al., 2009; Beach et al., 2010). Figure 3.5 shows the percentages willing to accept home monitoring of varying levels of intensity and sharing information with varying targets in order to prevent going to a nursing home. Respondents were less accepting of video monitoring—especially when done in the bedroom and bathroom—than sensors, even if this monitoring would prevent institutionalization. Respondents were also less accepting of sharing information with insurance companies than with family or a doctor even if it helped prevent institutionalization. These findings suggest that some privacy trade-offs may just "not be worth it" for a significant subset of baby boomers and older adults. Lastly, the survey found that more disabled and more educated respondents were more willing to make privacy trade-offs but that assistive device use (primarily wheelchair) was not related to willingness to make these trade-offs.

Lorenzen-Huber and colleagues (2010) have recently proposed a framework for privacy, technology, and aging that confirms and extends our work. These authors

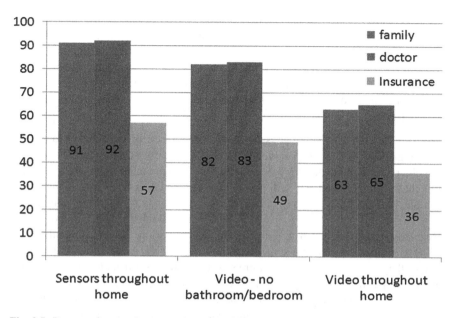

Fig. 3.5 Percent of respondents agreeing with different levels of home monitoring and sharing information with varying targets to prevent going to a nursing home

conducted focus group interviews with 65 older adults (age 70–85) who were allowed to observe and interact with various prototype technologies at Indiana University's Living Lab model home. The technologies ranged from the MD2 medication dispenser compliance system, to a *mirror motive* system that displays reminders and coordinates social engagement, to an *ambient plant* with sensors and lights to facilitate awareness between remote family members (eight total technologies). They found that older adult concerns about privacy were highly contextualized, individualized, and influenced by psychosocial motivations in late life. Factors that influenced perceived privacy included (1) the *perceived usefulness* of the technology (including awareness of their own perceived vulnerability or need for the technology), (2) *social relationships*, (3) *data granularity* (including level of granularity and data transparency), and (4) *sensitivity of activity*. Data granularity refers to the level of detail captured by the monitoring system—for example, full video versus obscured or ambient images from sensors. The researchers found that older adults were generally opposed to highly granular data collection like video, confirming our work cited above. Data transparency is concerned with what data are collected, stored, transmitted, or shared and with whom. Although they found that older adults were very naive about these issues, our work shows that concerns about privacy can vary when some of these parameters (what is recorded, how, and with whom it will be shared) are made clear. Their findings on sensitivity of activity—older adults were very concerned with being able to control the devices (i.e., turn them off), depending on what they were doing—mirror our findings that the acceptability of recording/sharing information differed by specific behavior (toileting and driving less acceptable). The Lorenzen-Huber et al. (2010) framework also points to the interrelationships between privacy concerns and other key issues for the acceptance and adoption of monitoring technologies by older adults, including perceived need for the technology by the older adult, perceived system usefulness, system demands, and the effects of monitoring technologies on social interaction and social connectedness. We conclude this section with a brief discussion of each of these issues.

Perceived Need A fundamental issue in the acceptance of monitoring (and all) technologies by older adults is perceived need. A recurring finding in the research with older adults and technology is that many of these potential users simply do not think they currently need, and will probably never need, this type of assistance. For example, Barrett (2008), analyzing data from the *Healthy @ Home* survey sponsored by AARP using national Web panels of 907 older adults (age 65+) and 1,023 caregivers, found that seven in ten older adults say various safety and health and wellness devices may not be something they need, while eight in ten caregivers thought they would have some or a great deal of difficulty persuading the people they help to use the technology. In a more detailed secondary analysis of the *Healthy @ Home* data, Schulz et al. (2010) found that believing a device "would not be something I need" was a significant (negative) predictor of possible use of health and wellness technology. In addition, believing that a safety technology "would make me look like I need help" was also predictive of lower likelihood of use. Lorenzen-Huber et al. (2010) also noted that their focus group participants tended to say that while *other* older adults would benefit from monitoring technology, they did not perceive a personal need. This may be the most fundamental barrier to the adoption of not only monitoring but other quality of life technologies by older adults, and it certainly deserves further study.

Perceived Usefulness Closely related to perceived need is the perceived usefulness of the technology. Older adults are more willing to adopt and use technology that they feel will be useful in meeting their needs. This is a key aspect of many technology uptake and acceptance models (e.g., Davis, 1989) and is certainly applicable to monitoring technology. Older adults are much more willing to be monitored if they perceive it as providing clear benefits.

System Demands Another key set of factors is system demands, including things like the complexity of the system and resulting difficulty in learning how to use it and maintain it. There is a large literature showing that older adults are generally less comfortable with and less often use technology than their younger counterparts (e.g., Czaja et al., 2006). Our own work showed that only 52% of our baby boomer and older adult registry survey participants were willing to use technology that required 5–10 h of training (vs. 81% for 2–3 h), or about 1 h of maintenance per day (56% vs. 94% for a few minutes of maintenance per day), even if the technology could help them perform ADL tasks (Schulz et al., 2010). We did find that those who were more disabled were more willing to make these trade-offs. However, these findings reinforce the need for simple, intuitive user interfaces that require minimal training of end users and other stakeholders. Another crucial system demand is the cost of the system. Multiple studies have shown that cost is perceived to be a major barrier to the potential adoption of technology for older adults and caregivers (e.g., Barrett, 2008; Schulz et al., 2010). However, a detailed discussion of this topic is beyond the scope of this chapter.

Social Connectedness The last issue we discuss is the impact of monitoring (and other) technologies for older adults on social interaction and connectedness. One of the most dramatic findings from our own work is that older adults are strongly opposed to systems or devices that reduce opportunities for social interaction, even if they would provide assistance with ADL or IADL tasks. Only 28% of the surveyed registry participants were willing to make such trade-offs (Schulz et al., 2010). Lorenzen-Huber et al. (2010) also found that the older adults in their focus groups were very concerned about monitoring technology intruding on their adult children's lives, that technology should never replace human contact, and that they should remain full participants in any exchange of data with family or healthcare providers. Finally, as noted in the introduction to this section, there are monitoring technologies that focus on maintaining and increasing *social connectedness*. All of this work reflects the fundamental desire for many older adults to remain socially integrated and connected and that technology that reduces these opportunities is not likely to be acceptable. Clearly there is a need for more research in this interesting area which has broad implications for a wide range of quality of life technologies.

3.5   Issues Regarding Stress with, Training for, and Acceptance of Healthcare Technology

As previously discussed, for healthcare technology to be useful and usable for older populations, the following issues need consideration: stress associated with initial use, training for use and for maintenance, and barriers to acceptance of technologies into a daily routine. As an example, consider the case of an older adult diagnosed with Type II diabetes, something that is occurring with increasing frequency in the population (Boyle et al. 2001; Mainous et al., 2007). The person is instructed by their physician to change their diet, to increase their exercise levels, and, in extreme cases, to use a glucose meter to monitor blood sugar levels. When first told about the diagnosis, how likely are they to understand and remember what they are being told? What kind of instruction will facilitate proper adherence to the treatment regimen? Finally, what are the barriers to integrating the treatment regimen into a daily routine? Also, who will teach them to use and maintain the glucose meter? As discussed in the following section, these issues can be very stress inducing which can ultimately interfere with a person's willingness or ability to engage in a new task or interact with a new type of technology.

Stress Classic models of stress and coping (e.g., Folkman, 1997; Lazarus & Folkman, 1984) emphasize the importance of the cognitive appraisal process in judging whether a situation will be perceived as stressful or not. A useful way to conceptualize stress effects is to consider the degree of match between the demands on the individual and the resources that they believe that they can bring to the task (e.g., Figure 1 in Charness, 2010). If the demands are in balance with the perceived resources, then there will be little experience of stress. If there is concern that demands are greater than resources, stress will be experienced in the form of physiological arousal (elevated blood pressure, heart rate) as well as in cognitive changes (increase in negative emotion and possibly increased workload when performing the task if someone is worried about their performance).

Importantly, long-term stress is associated with negative changes in health. For instance, the stress of caregiving for someone with dementia may result in increased morbidity and mortality (Schulz & Beach, 1999). But even short-term stress can interfere with someone's ability to perform a task correctly and may be differentially harmful to older adults. Stress can narrow the field of attention, making it more difficult for individuals to cope with the demands of a complex task. Not surprisingly, new technology, including healthcare technology, can evoke stress in a user, particularly when that user is an older adult unfamiliar with the technology. Though older adults are not averse to learning to use new technology if it is perceived to be usable and useful (Mitzner et al., 2010), they are less likely to use it compared to younger adults (Czaja et al., 2006) and more likely to experience stress when using it (Sharit & Czaja, 1994).

There is also evidence that older adults are more likely to experience an elevated physiological response to stressors than younger adults, at least when the outcome measures include heart rate, pulse, and blood pressure (Uchino, Berg, Smith, Pearce, & Skinner, 2006; Uchino, Holt-Lunstad, Bloor, & Campo, 2005), though this is not always the case (Dijkstra, Charness, Yordon, & Fox, 2009). A potential solution to older adult sensitivity to stressors, including the introduction of new healthcare technology, is to provide them with time to settle in with a new procedure and, more importantly, to provide them with the training and practice necessary to use the device or system properly. They also need access to technical support. Anxiety associated with a previously unknown situation may dissipate with appropriate education about the situation and some practice with the new task or device. It is also important to reassure the person that they will be capable of mastering the new activities and to provide sufficient time for them to learn. One possibility to consider is demonstrating device use with an age-matched model.

Training Unfortunately, much training for use of health-related technology is *just in time* training. That is, a device is introduced (prescribed) and the senior must immediately learn to use it without any further instruction or support. This type of just in time training can be quite stressful and lower a person's belief that they will be able to eventually use the device or system irrespective of the person's age. Take the case of someone newly diagnosed with diabetes. The emotional reaction that they have to the diagnosis is stressful by itself and can be expected to interfere with their ability to attend to the healthcare professional's instructions about changing diet and exercise levels. At the same time this person has to think about accessing and using a new device, a blood glucose meter. Aside from an expected narrowing of perceptual and cognitive focus, it is also likely that repetitive or ruminative thoughts (Watkins, 2008) will reduce the cognitive resources available to process any instructions being given soon after the diagnosis. (This reaction can be conceptualized within the stress framework described above.)

So, training techniques should incorporate ways to reduce any emotional distress that the senior may be experiencing, possibly by training relaxation techniques (e.g., Dijkstra et al., 2009). Another factor worth considering is that older adults learn more slowly than younger adults (e.g., Charness, Kelley, Bosman, & Mottram, 2001), so sound training practice should take the slower learning rate into account, possibly by arranging for self-paced training. It is not yet clear what the best training methods are for older adults (e.g., Charness & Czaja, 2006) in part because few studies have investigated the upper end of the life span (those most vulnerable to acute and chronic health conditions), with most being concerned with older worker populations (e.g., Callahan, Kiker, & Cross, 2003). Aside from self-paced learning, learning in small groups seems to be differentially beneficial to older adults. However, training that used to be provided directly by the healthcare professional is increasingly being off-loaded to the client, via instruction sheets, manuals, and even video sources. It is a safe bet that such instructional materials will be distributed more frequently via the Internet. However, current cohorts of older adults are among the least likely users of the Internet, with only about 42% of those age 65 and over in the United States report using the Internet in the past year (Pew Internet and American Life Project, 2008).

Acceptance and Maintenance There are a number of macroscale factors influencing acceptance of health technology. Models of technology acceptance (e.g., the TAM model of Venkatesh & Davis, 2000) stress that acceptance of a new technology in a work environment depends on a potential user's beliefs about how useful the technology will be as well as on how easy to use it appears to be. This model can be applied to health technology with some minor modifications. The perceived importance of the device for health probably varies more than the perceived importance of a device for work (where you may have the option of leaving your job environment or even deciding not to adopt the device). You are more willing to use something if you are likely to die without it, for instance, a glucose meter for a Type 1 diabetic. Another factor is the cost of the device (money, time invested in use). In the absence of societal/governmental support for purchase, it may be too costly for an individual to purchase or lease a device. As in the technology acceptance model (TAM), a critical factor is its usability, how easily or comfortably the device can be deployed, used, and maintained. For instance, for someone with mild sleep apnea, a continuous positive airway pressure device may be so unwieldy and interfere so much with sleep initially that its use will be discontinued. Another factor that is not often considered (except perhaps by marketers) in acceptance is the aesthetics of the device, including any stigma associated with its use.

Obtrusiveness has been specifically identified as a potentially important factor in home telehealth (Hensel, Demiris, & Courtney, 2006).

Given that a device has been purchased, how likely is the user to persist with its use? One important factor is the user's attitudes toward the device. Models such as the theory of planned behavior (Ajzen & Fishbein, 1977) postulate that behavioral intentions follow from attitudes toward the behavior, subjective norms about the behavior (e.g., the opinions of family and friends), willingness to comply with those norms, as well as beliefs about one's ability to engage in the behavior (e.g., self-efficacy: Bandura, 1982). So, assume that a senior has been given a diagnosis and instruction on how to manage the disease process (e.g., for a chronic condition such as adult-onset diabetes). How likely is it that the senior will persist with the suggested routine? Of the factors in the Ajzen and Fishbein model, subjective norms and self-efficacy seem likely to play the biggest roles. For instance, subjective norms could be important in adhering to diet. If friends go along with the idea that a diabetic can cheat a little on desserts, they may do so at the party. The assistive device literature has identified both positive and negative factors in discontinuing use of a device. A summary with minor changes can be seen in Table 3.1, reclassified by perceived usefulness, usability factors.

Gitlin (1995) outlined some of the factors that have been identified for discontinuance by older adults. Factors include (1) improvement in the person's capabilities so that the device is unnecessary, (2) inability to use because the device depended on another device that has been discontinued, (3) lack of knowledge of how to use the device, (4) poor fit for the device to the user's environment, and (5) the device broke or was lost. For example, a complex telehealth device (such as videoconferencing equipment) is likely to fall prey to inter-device dependence issues and lack of knowledge problems. What the work on acceptance points out is that even if a device is initially perceived as useful and usable, and can be obtained at a reasonable cost, changing abilities and attitudes are potential barriers to incorporating health technology into a daily routine.

3.6   Conclusions

The growth in the number of older people, especially those who represent the oldest old, has vast implications for the healthcare system given the increased propensity towards illness and disability with advancing age. As discussed throughout this chapter, healthcare technologies hold promise in enhancing the ability of older adults and their families to access needed healthcare information and services and ultimately the ability of older people to live independently in the community. For example, home-based monitoring technologies offer numerous opportunities for increasing the safety, health and wellness, and social connectedness of older adults. These technologies also offer many benefits to healthcare providers and family caregivers. Internet-based health applications also provide opportunities for older adults and family members to be more involved in health self-management and to have greater access to resources and information and social support.

Table 3.1 Positive and negative factors in assistive device discontinuance

Negative factors in discontinuance	Positive factors in discontinuance
Usefulness	Usefulness
Never used or installed	Increased function makes it unnecessary
Seen as unnecessary	Replaced by an alternative solution
Negative views of device	Usability
Depression	Replaced by personal assistance
Failure to acknowledge/accept disability	Replaced by better equipment
Device selected without adequate consultation	
Lost device	
Usability	
Decreased function makes it unusable	
Device too difficult to use (size, weight, energy consumption)	
Unsafe to use	
Poor aesthetics	
Stigma	
Overly complex instructions	
Lengthy setup time	
Malfunction or failure of device	
Pain/discomfort in use	
Maintenance cost	
Device damages property	
Wrong device obtained	
Lack of sufficient training	
Difficulty accessing device	
Device requires personal assistance	
Device depends on another device	

Generally older people are receptive towards using these technologies especially if use of these devices, systems, and applications increases their potential to remain independent. However, as pointed out in our discussion, technology uptake is not ubiquitous among older people. There are a number of existing barriers to widespread adoption for current cohorts of older adults including large and diverse user groups with varying needs and abilities, lack of knowledge about the potential benefits of technology; low technology self-efficacy and anxiety; cost and accessibility; training opportunities; and system design characteristics. Many older adults are unaware of existing technologies, the benefits these technologies have to offer, and how to access these devices and applications. Further, many people do not have access to training and technical support and have some anxiety about their ability to use and maintain technology systems. Other concerns are related to issues regarding privacy, security, and reliability. Finally and importantly, system designers need to be aware of existing design guidelines for older users and adopt a user-centered design approach where older adults and family caregivers are actively involved in the design process. Clearly there is a need for more research in this area to answer questions regarding optimal system designs, training strategies, implementation processes, and cost-effectiveness issues.

Currently, the potential for technology to enhance the well-being and independence of older adults is not being actualized. For the full potential of technology to be realized for older adults and their families, the needs and abilities of older people must be considered in the design of technology systems and the design of implementation and training strategies. Unless older adults have full access to these healthcare technologies, they will be at an increased disadvantage in today's technology-oriented healthcare environment and the potential for age-related healthcare disparities will be increased.

References

Ajzen, I., & Fishbein, M. (1977). Attitude-behavior relations: A theoretical analysis and review of empirical research. *Psychological Bulletin, 84,* 888–918.

Alwan, M., Wiley, D., & Nobel, J. (2007). *State of technology in aging services.* Washington, DC: Center for Aging Services Technologies (CAST).

Alzheimer's Association. (2010). *Alzheimer's disease facts and figures.* Chicago, IL: Author.

Bandura, A. (1982). Self-efficacy mechanisms in human agency. *American Psychologist, 37,* 122–147.

Barrett, L. L. (2008). *Healthy@home research report.* Washington, DC: AARP Foundation.

Beach, S. R., Schulz, R., Downs, J., Matthews, J., Barron, B., & Seelman, K. (2009). Disability, age, and informational privacy attitudes in quality of life technology applications: Results from a national web survey. *Transactions on Accessible Computing (TACCESS), Special Issue on Aging and Information Technologies, 2(1),* Article 5.

Beach, S., Schulz, R., Downs, J., Matthews, J., Seelman, K., Barron, B., et al. (2009). *End-user perspectives on privacy and other trade-offs in acceptance of quality of life technologies.* Paper presented at the First International Symposium on Quality of Life Technology, Pittsburgh, PA.

Beach, S., Schulz, R., Downs, J., Matthews, J., Seelman, K., Person Mecca, L., et al. (2010). Monitoring and privacy issues in quality of life technology applications. *Gerontechnology, 9(2),* 78–79.

Boyle, J. P., Honeycutt, A. A., Venkat Narayan, V. M., Hoerger, T. J., Giess, L. S., Chen, H., et al. (2001). Projection of diabetes burden through 2050: Impact of changing demography and disease prevalence in the U.S. *Diabetes Care, 24,* 1936–1940.

Callahan, J. S., Kiker, D. S., & Cross, T. (2003). Does method matter? A meta-analysis of the effects of training method on older learner training performance. *Journal of Management, 29,* 663–680.

Cantor, N., & Sanderson, C. (1999). Life task participation and well-being: the importance of taking part in daily life. In D. Kahneman, E. Diener, & N. Schward (Eds.), *Well-being: The foundation of hedonic psychology* (pp. 230–243). New York: Russell Sage Foundation.

Center for Disease Control and Prevention. (2010). *Improving and extending quality of life among older Americans: At a glance 2010.* Retrieved from http://www.cdc.gov/chronicdisease/resources/publications/AAG/aging.htm

Charness, N. (2010). The health care challenge: Matching care to people in their home environments. In S. Olson (Rapporteur.), *The role of human factors in home health care: A workshop summary* (pp. 73–116). Washington, DC: National Research Council of the National Academies, National Academies Press.

Charness, N., & Czaja, S. J. (2006). *Older worker training: What we know and don't know.* AARP Public Policy Institute, #2006-22. Retrieved from http://www.aarp.org/research/work/issues/2006_22_worker.html

Charness, N., Kelley, C. L., Bosman, E. A., & Mottram, M. (2001). Word processing training and retraining: Effects of adult age, experience, and interface. *Psychology and Aging, 16,* 110–127.

Cobb, S. (1976). Social support as a moderator of life stress. *Psychosomatic Medicine, 38*, 300–314.

Czaja, S. J., Charness, N., Fisk, A. D., Hertzog, C., Nair, S. N., Rogers, W. A., et al. (2006). Factors predicting the use of technology: Findings from the Center for Research and Education on Aging and Technology Enhancement (CREATE). *Psychology and Aging, 21*, 333–352.

Czaja, S. J., Sharit, J., Hernandez, M. A., Nair, S. N., & Loewenstein, D. (2010). Variability among older adults in Internet Health information–seeking performance. *Gerontechnology, 9*, 46–55.

Czaja, S. J., Sharit, J., & Nair, S. N. (2008). Usability of the Medicare health web site. *Journal of the American Medical Association, 300*(7), 790–791.

Davis, F. (1989). Perceived usefulness, perceived ease of use, and user acceptance of information technology. *MIS Quarterly, 13*, 319–339.

Dijkstra, K., Charness, N., Yordon, R., & Fox, M. (2009). Changes in physiological and self-reported stress in younger and older adults after exposure to a stressful task. *Aging, Neuropsychology and Cognition, 16*, 338–356.

Dykstra, P. (1995). Loneliness among the never and formerly married: The importance of support-ive friendships and a desire for independence. *Journal of Gerontology: Psychological Sciences and Social Sciences, 50B*, S321–S329.

Ellaway, A., Wood, S., & MacIntyire, S. (1999). Some to talk to? The role of loneliness as a factor in the frequency of GP consultations. *British Journal of General Practice, 49*, 363–637.

Ellis, P., & Hickie, I. (2001). What causes mental illness. In S. Bloch & B. Singh (Eds.), *Foundations of clinical psychiatry* (pp. 43–62). Melbourne: Melbourne University Press.

Folkman, S. (1997). Positive psychological states and coping with severe stress. *Social Science & Medicine, 45*, 1207–1221.

Fox, S. (2011a). *Health topics*. Pew Internet & American Life Project. Retrieved from http://www.pewinternet.org/Reports/2011/HealthTopics.aspx

Fox, S. (2011b). *Peer-to-peer healthcare*. Pew Internet & American Life Project. Retrieved from http://www.pewinternet.org/Reports/2011/HealthTopics.aspx

Fox, S. & Purcell, K. (2010). *Chronic disease and the Internet*. Pew Internet & American Life Project. Retrieved from http://www.pewinternet.org/Reports/2010/Chronic-Disease.aspx.

Gitlin, L. N. (1995). Why older people accept or reject assistive technology. *Generations, 19*, 41–47.

Hensel, B. K., Demiris, G., & Courtney, K. L. (2006). Defining obtrusiveness in home telehealth technologies: A conceptual framework. *Journal of the American Informatics Association, 4*, 428–431.

Kinsella, K., & He, W. (2009). *An aging world 2008: International population reports*. Retrieved from http://www.census.gov/prod/2009pubs/p95-09-1.pdf

Lazarus, R. S., & Folkman, S. (1984). *Stress, appraisal, and coping*. New York, NY: Springer.

Lenhart, A. (2010). *Cell phones and American adults*. Pew Internet & American Life Project. Retrieved from http://pewinternet.org/Reports/2010/Cell-Phones-and-American-Adults.aspx

Lorenzen-Huber, L., Boutain, M., Camp, L. J., Shankar, K., & Connelly, K. H. (2010). Privacy, technology, and aging: A proposed framework. *Ageing International, 36*(2), 232–252.

Mainous, A. G., Baker, R., Koopman, R. J., Saxena, S., Diaz, V. A., Everett, C. J., et al. (2007). Impact of the population at risk of diabetes on projections of diabetes burden in the United States: An epidemic on the way. *Diabetalogia, 50*, 934–940.

Mitzner, T. L., Boron, J. B., Fausset, C. B., Adams, A. E., Charness, N., Czaja, S. J., et al. (2010). Older adults talk technology: Technology usage and attitudes. *Computers in Human Behavior, 26*(6), 1710–1721. doi:10.1016/j.chb.2010.06.020.

National Alliance for Caregiving. (2009). *Caregiving in the U.S: A focused look at those caring for someone age 50 or older*. Retrieved from http://www.caregiving.org/data/FINALRegular ExSum50plus.pdf

Pewinternet. (2008). *Statistics pages*. Retrieved from http://www.pewinternet.org/Static-Pages/Trend-Data/Whos-Online.aspx

Schulz, R., & Beach, S. (1999). Caregiving as a risk factor for mortality: The caregiver health effects study. *Journal of the American Medical Association, 282*, 2215–2219.

Schulz, R., Beach, S., Matthews, J., Seelman, K., Person Mecca, L., & Courtney, K. (2010). Design preferences for technologies that enhance functioning among older and disabled individuals. *Gerontechnology, 9*(2), 79–80.

Sharit, J., & Czaja, S. J. (1994). Ageing, computer-based task performance, and stress: Issues and challenges. *Ergonomics, 37*, 559–577.

Sharit, J., Hernandez, M., Czaja, S. J., & Pirolli, P. (2008). Investigating the roles of knowledge and cognitive abilities in older adult information seeking on the web. *ACM Transactions on Computer-Human Interaction, 15*, Article 3.

Skubic, M., Alexander, G., Popescu, M., Rantz, M., & Keller, J. (2009). A smart home application to eldercare: Current status and lessons learned. *Technology and Health Care, 17*, 183–201.

Smith, A. (2010). *Home Broadband 2010*. Pew Internet & American Life Project. Retrieved from http://www.pewinternet.org/Reports/2010/Chronic-Disease.aspx

Taha, J., Czaja, S. J., & Sharit, J. (2009). Use of and satisfaction with sources of health information among older Internet users and non-users. *The Gerontologist, 49*, 663–673.

U.S. Census Bureau. (2011). *Information and communications*. Statistical Abstract of the United States: 2011. Retrieved from http://www.census.gov/prod/2011pubs/11statab/infocomm.pdf

U.S. Department of Health and Human Services. (2011a). *Health.Gov*. Retrieved from http://www.hhs.gov

U.S. Department of Health and Human Services. (2011b). *Nationwide health information network*. Retrieved from http://healthit.hhs.gov/portal/server.pt?open=512&objID=1142&parentname=CommunityPage&parentid=4&mode=2

Uchino, B. N., Berg, C. A., Smith, T. W., Pearce, G., & Skinner, M. (2006). Age-related differences in ambulatory blood pressure during daily stress: Evidence for greater blood pressure reactivity with age. *Psychology and Aging, 21*, 231–239.

Uchino, B. N., Holt-Lunstad, J., Bloor, L. E., & Campo, R. A. (2005). Aging and cardiovascular reactivity to stress: Longitudinal evidence for changes in stress reactivity. *Psychology and Aging, 20*, 134–143.

Venkatesh, V., & Davis, F. D. (2000). A theoretical extension of the technology acceptance model: Four longitudinal case studies. *Management Science, 46*, 186–204.

Watkins, E. R. (2008). Constructive and unconstructive repetitive thought. *Psychological Bulletin, 134*, 163–206.

Zaphiris, P., Ghiawadwala, M., & Mughal, S. (2005). *Age-centered research-based web design guidelines*. CHI '05: CHI '05 Extended Abstracts on Human Factors in Computing Systems, Portland, OR, USA, 1897–1900.

Chapter 4
Assistive Technologies: Ethical Practice, Ethical Research, and Quality of Life

Andrew Eccles, Leela Damodaran, Wendy Olphert, Irene Hardill, and Mary Gilhooly

4.1 Introduction

Much has been written about the benefits–for example, greater independence, autonomy, and dignity–which can derive from the use of assistive technologies with older people (Loader, Hardey, & Keeble, 2009; McCreadie & Tinker, 2005; Pols & Moser 2009). These benefits have been well researched, clearly expressed in the literature, and remain uncontested here. Apart from benefits to individuals and carers, assistive technologies can release funds for other applications, and where this is in care settings funded through public expenditure, the cost savings that might arise from their application may afford the opportunity for more effective targeting of taxpayers' resources. As the Audit Commission, the guardian of public expenditure in the United Kingdom (UK) noted, the use of technology represents the unusual possibility of providing cost savings at the same time as better service provision

A. Eccles, M.Phil. (✉)
School of Applied Social Sciences, University of Strathclyde,
Glasgow G4 OLT, Scotland
e-mail: andrew.eccles@strath.ac.uk

L. Damodaran, Ph.D. • W. Olphert, Ph.D.
Information Technology and Social Research Group, Department of Information Science,
Loughborough University, Holywell Park, Loughborough LE113TU, UK
e-mail: l.damodaran@lboro.ac.uk; c.w.olphert@lboro.ac.uk

I. Hardill, Ph.D.
Department of Social Sciences, Northumbria University, Lipman Building,
Newcastle upon Tyne NE1 8ST, UK
e-mail: Irene.hardill@northumbria.ac.uk

M. Gilhooly, Ph.D.
Institute for Ageing Studies, Brunel University, Uxbridge, Middlesex, UB8 3PH, UK
e-mail: Mary.Gilhooly@brunel.ac.uk

A. Sixsmith and G. Gutman (eds.), *Technologies for Active Aging*,
International Perspectives on Aging, DOI 10.1007/978-1-4419-8348-0_4,
© Springer Science+Business Media New York 2013

(Audit Commission, 2004). The merits then are evident. But there is a need also to be alert to the ethical questions that arise as a concomitant to the use of new technologies and to address what Mort, Roberts, and Milligan (2009) have argued is an "ethical and democratic deficit in this field which has arisen due to a proliferation in research and development of advanced care technologies that has not been accompanied by sufficient consideration of their social context" (p. 85). This chapter will raise these ethical issues, alight on the potential deficits, and highlight some of the policy and practical issues that might warrant further inquiry. It does so by addressing three key areas. First, it considers ethical approaches commonly in use and their limitations for application in the field of assistive technologies. Second, it explores the ethical issues that arise around the design and execution of research with users of assistive technologies. Third, it raises the question of whether or not assistive technologies contribute to a better quality of life (QoL) for recipients, not least because QoL is explicitly included as an intended policy outcome of the deployment of such technologies (Scottish Government, 2009; Telecare Services Association, 2010). The chapter draws its examples primarily from the experience of the UK (and its devolved polities), where the policy objectives of governments for extending the use of assistive technologies are particularly ambitious. We are mindful of the methodological pitfalls of cross-national applicability, but the discussion should have broader resonance, as ethical approaches and practices come to terms with a technologically fast-changing world. Our use of the term assistive technologies embraces the definition of any item, piece of equipment, product, or system, whether acquired commercially off the shelf, modified, or customized, that is used to increase, maintain, or improve the functional capabilities of individuals with disabilities (Technology-Related Assistance for Individuals with Disabilities Act, 1988 P.L.100/407); more practicably, as Cowan and Turner Smith (cited in McCreadie & Tinker, 2005) note, it refers to any device or system that allows an individual to perform a task that they would otherwise be unable to do or increases the ease and safety with which the task can be performed. We note in particular that some of the more interesting ethical challenges have arrived with the advent of telecare sensor-based technologies, which open up important issues around privacy, autonomy, and the potential for replacement of human care through remote monitoring.

4.2 The Policy and Practice Context

The policy drive toward the use of assistive technologies, particularly for the care and well-being of older people, has developed rapidly in recent years. In the United Kingdom, since 2005, there have been government programs for telecare development, with a similar program of telemedicine technologies being implemented in the delivery of health care in the community. Telecare policy now sits at the strategic heart of the delivery of care services. Care policy in the UK is a complex mix of family input, use of the private sector, voluntary organizations, and the State (see Phillips, 2007 for a useful overview). Of these, the State has long played the key role in health care provision and albeit in more complex ways in terms of funding, a major role in social care. This role played by the State in the UK should also be seen in the context of comparatively (certainly in European terms) low levels of obligation placed on family members to take responsibility, either in a legal or cultural sense, for their aging parents. Thus governments, straddled with costs but also seeing opportunities for innovative forms of care for older people are pushing the gerontechnology agenda. The telecare development program of the Scottish Government (2008), for example, proposes that by the year 2015, "all new homes, public and private, and all refurbished social housing, will be fitted with the capacity for care and health services to be provided interactively via broadband from day one of occupation (and) remote long term condition monitoring undertaken from home will be the norm" (p. 6). This policy is explicitly linked to demographic change and the rise in numbers of older people, particularly those over the age of 75, relative to the population as a whole. The nexus between an older population that may need care and a workforce able to provide this care and fund it through taxation has been expressed as the *dependency ratio* (European Union Public Health Information, 2009). A policy discourse has developed around this ratio, in which major technology providers (e.g., Tunstall, 2009) note an impending *demographic time bomb*, to the extent that this discourse readily assumes the *necessity* of technological solutions for future service delivery. This is contestable territory. To have an explicit government policy objective that telecare services grow as quickly as possible (Scottish Government, 2008) indicates that the role to be played by these technologies is already beyond debate. But there are important ethical issues raised by the increased use of assistive technologies which do indeed need the space to be debated.

4.3   Ethical Frameworks

The ethical angle is often absent or of limited import in policy discussions around increased use of assistive technologies. This might be explained by the fact that different professions engaging in human services already have ethical codes of practice (although the codes themselves are then open to interpretation). But are these codes enough to deal with rapid and innovative technological change? A review of the ethical frameworks currently employed by the various agencies engaged in the provision of assistive technologies suggests they are limited in scope. This, in part, reflects the practical reality that ethical frameworks have to be understood by practitioners and their terminology has to resonate with the care assessment process (Bouma, 2010). In the UK context, the practice of health and social care is additionally molded by a broader canvas of managerialism and performance targets (Meagher & Parton, 2004), factors which may sit awkwardly with the imprecision and uncertainty that ethical questions pose. The key framework in widest use is grounded in a biomedical approach and rests on four key principles: autonomy, beneficence, non-maleficence, and justice (Beauchamp & Childress, 2001). These are profoundly important concepts, but they need to be tempered by a broader range of ethical enquiry and contextual understanding, for example, the way in which they will be understood across different cultural settings or different professions. Indeed, the essentially medical nature of this approach may suit aspects of health technology, for example, telemedicine, but sit more uneasily in social care contexts. The biomedical approach usefully illuminates ethical issues about medical interventions at specific junctures in people's lives but the ongoing, and often shifting, needs of individuals in domiciliary care require a more subtle ethical enquiry. For example, what Wilmot (1997) calls the primacy of autonomy sits at the heart of much of the assistive technologies agenda. Unpacking arguments around autonomy is rarely straightforward, and the importance placed on autonomy may underplay the significance of our interdependence with each other. Furthermore, the ASTRID framework (Frisby, 2000), drafted primarily for dementia care, notes that greater independence arising from autonomy might also bring with it greater isolation, what Wilmot would term unwanted autonomy. While there is clear evidence that the use of technology, not least communication technology and the development of virtual communities, can actually alleviate isolation among older people (Blaschke, Freddolino, & Mullen, 2009), where there is the potential for technology to increase isolation there lies the concomitant risk of higher instances of depression (see Lowe, 2009). Although the notion of autonomy has a central place in discussions around ethical practice, it takes on different meanings for different groups of people. For example, in societies where there is a Confucian tradition, the full worth of being autonomous is only recognized in relation to a more complex array of interdependence and reciprocities with others (Tao & Drover, 1997). Thus there needs to be space in the consideration around the use of assistive technologies for their culturally sensitive application, especially given that the biomedical four principles that underpin the ethical frameworks are in such common use. More evidence on how these assessment decisions, balancing care, risk, and potential harm, are in fact calculated in relation to assistive technologies is needed. As Hanson, Osipovic, and Percival (2009) note in their study of the impact of sensor-based technology: "In order to make 'sense of sensors' alongside the data provided by the devices, one needs rich contextual information that is normally accumulated through social interactions between caregivers and care receivers, a two-way communication process that can best be described as a 'dialogue of care'" (p. 111).

4.4   Alternative Ethical Approaches

The argument here is that the predominant ethical framework in use around the implementation of, for example, telecare technologies serves an important but limited function. That it remains essentially a biomedical framework but is often being used in the context of social care prompts the need for other sources of ethical enquiry. This does not necessarily mean that the already complex task of assessment for the use of assistive technologies should be subject to a further accretion of ethical codes and guidelines, but a broader range of ethical thinking should be an important part of policy formation and consideration of policy implementation around the role of assistive technologies in health and care services with older people.

One such alternative ethical approach comes from the tradition of an ethic of care (Tronto, 1994). This is particularly relevant to community-based health and social care, where assistive technologies are at the forefront of new ways of delivering care services, as it is here that interaction between professionals and service users may exist on relationships developed and sustained over a longer period than in the acute medical settings that inform the biomedical approach. Barnes (2006) notes the way in which social care workers often go beyond *tasks* to develop *relationships* over and above contractual obligations, relational approaches to care that are contextual and not necessarily rule based nor uniform in application. Some older people value independence highly and might regard being the recipient of care inconvenient, at best, and indeed potentially demeaning. Others, for example, people who are socially isolated after they have lost lifelong partners, might welcome human intervention. This relational aspect to care may thus be played out quite differently in different settings. It may also, adversely, prompt reluctance by social care professionals to engage with technology and its possibilities for the recipients of care services. In emphasizing the importance of relationships in human services, we need to guard against assumptions that technologically based care is axiomatically inferior to care based on human relationships. As Pols and Moser (2009) argue "in discussions about the use of new technologies in health care, including the most recent versions appearing as telecare, there is the fear that cold technologies will be implemented at the cost of warm human care" (p. 160). However, they conclude from their research that "the opposition between cold technology and warm care does not hold, but that there are different relations between people and technologies within different use practices allowing different affective and social relations" (p. 159). So, again, it is the *specific context* in which decisions are made that is crucial. There is a risk in polarizing this discussion around human delivery of care services and technological provision. Human care services have historically been under fiscal pressure and have often denied recipients much real choice in delivery. Indeed recent evidence from the inquiry into home care provision in the UK by the Equality and Human Rights Commission paints a bleak picture of often impersonal, very time-limited, and inconsistent care delivery in which the potential for relationships to develop between carers and older people appears to be increasingly remote (Equality and Human Rights Commission [EHCR], 2011). Holding this up as an inherently better model of care delivery, when technology might enable some tasks to be undertaken more reliably, is illogical. Of course many of the care tasks discussed in the EHCR report cannot easily be replaced by technology; nonetheless, especially as fiscal pressures on care delivery for older people increase, we should be wary of assuming that human care services as currently configured are axiomatically somehow better. That said, where technologies can provide for care needs, they will be operating in a social care agenda where greater choice is the new mantra; but if service users do not want these technologies, is this a genuine choice which can be upheld at a time when households are increasingly being equipped with connectivity as part of government strategy (Scottish Government, 2008) on meeting future health and social care needs?

4.5 Remote Monitoring and Decision-Making

Further ethical issues arise beyond the immediate environment of technology use. The remote monitoring of service users–for example, through monitoring using home-based sensors–raises questions about the response to an alert in the control center which oversees the monitored spaces. Straightforwardly, decisions on how to triage alerts for intervention can be based on users' existing medical histories; thus, an alert from someone with a known heart condition might take priority over other signals for help. Beyond this, we need to understand how judgments based on responding to the demands of remote sensors are made. Intuitionism (Driver, 2007) offers up lines for reflection here. Human beings have intuitive responses to right and wrong courses of action in the face of immediate human dilemmas that are not based on calculation or recourse to abstract concepts. But are the care needs of service users who are monitored remotely perceived in the same way as they would be if there was immediate human involvement? Will decisions about a course of action be different when the immediacy of care needs is filtered through a process of remote monitoring and subsequent triaging? There is research from the use of telemedicine (Finch, Mort, Mair, & May, 2008) which suggests that human engagement involves sensitivity to user conditions that technology may not pick up remotely. In the social care context, for example, the replacement of a brief early morning visit by domiciliary carers with a remotely based telephone inquiry about clients' well-being might elicit the response from a service user that everything is fine. Feeling fine might be a culturally influenced response of not wanting to make a fuss rather than an accurate account of actual circumstances. However, domiciliary carers on home visits may intuitively sense when a clients' circumstances are not ideal, especially if they have developed a relationship with clients over time. Thus do remote decision-making processes make a difference to the quality of care and by what calculation do we maintain a supply of care workers for periods when remote monitoring is insufficient to the task of addressing care needs? These are particularly relevant issues in the context of older people, where conditions of health may alter rapidly in a short space of time. Equally, how robust, in the highly pressured world of service delivery, are the review processes that ensure equipment is being used properly and is still fit for the purpose it held at the time of installation?

4.6   The Virtuous Practitioner

Such is the complexity of factors–risk, protection, and empowerment–that attend the use of these technologies that frameworks of ethical practice in themselves may be of limited practical use and the *virtues* associated with working in the field of human services may have to come to the fore (Banks & Gallagher, 2009). This approach would link awareness of ethical codes and frameworks (which, in practice, are variously interpreted and variously employed) to the essential virtuousness of practitioners through their professional training and vocation, vocation being the underpinning sense that an individual might have about why they feel suited and committed to the work they do (see Cooper, 2012, for further discussion of this). Thus the difficulties of interpreting contexts for the use of technology might be less problematic in the presence of the virtuous practitioner, who might be expected to take the morally sound course of action. This might prove more difficult when it comes to assistive technologies, however, unless there is a highly developed common understanding of what constitutes a virtuous approach in relation to the use of technology in care settings. Equally, the increasingly fluid world of assessing for care needs across professional boundaries (where there may be different ethical codes, or at least understanding of these codes) may mean that recourse to virtue per se by dint of professional training or vocational calling is not a given. For example, in the UK, the greater incidence of interprofessional working has seen the development of shared assessment tools for the assessment of health and care needs. These assessment tools now include sections for consideration of the use of assistive technology-based solutions to these needs. The thrust of recent policy has been to assume that common datasets are sufficiently straightforward to collect, such that inconsistencies across professional disciplines will be minimal. However, research on shared assessment (Eccles, 2008) has noted inconsistencies in assessment, for example, understanding of consent, obtaining agreement from service users for information to be shared and in the quality of the narrative element of assessment (which might explore the social context in which technologies might be employed). The interest here, then, is in the consistency of assessment that recommends the use of assistive technology. Professional attitudes to care, ethical frameworks, and the use of these frameworks and vested interests in maintaining the status quo all suggest inconsistency of approach. This is not, in itself, surprising as the impact of professional domains has long been recognized (Irvine, Kerridge, McPhee, & Freeman, 2002) but how assessments are made and how older people have their care managed in the context of the use of technology across these different domains would merit further enquiry. Decision-making may well be virtuous, but consistency of understanding of what constitutes virtue is open to question, as is just how virtuous decision-making *can* be, not only across professions but in the wider context of a culture of different performance indicators across these professions (Loader, 2005). As previously suggested, there is a deficit around our knowledge of how ethical considerations play out in the course of technology-based health and social care practice. The provision of telecare technologies undoubtedly has the potential to offer significant benefits to individuals. However, the process through which informed consent to, for example, monitoring and surveillance is gained routinely highlights the inadequacies of the biomedical model in its neglect of social and behavioral aspects. Although emphasis is given to gaining informed consent, there are few tools and techniques in use to help people understand the far-reaching implications of surveillance (e.g., the fact that complete strangers in remote control rooms may be observing the individual's behavior in his/her own home). The significant potential benefits offered by such technologies are not in question, but the threat of radically changing the character of the home from being a space which has been traditionally regarded as safe, secure, and private (Twigg, 1999) does not appear to be acknowledged or evaluated in the context of existing models of telecare delivery. As a consequence, and in the absence of any alternative model of practice, the provision of information, meaningful explanation, and opportunities for gaining understanding and learning about the implications of implementation of telecare technologies do not appear to be an essential part of the process of gaining informed consent to the presence and use of telecare in the home. Furthermore, it does not appear to be part of the contractual arrangements between telecare/telehealth providers and customers (in the UK context), yet consent forms are routinely signed by end users (often vulnerable older and disabled people) without these issues being explained or discussed, nor without the opportunity to develop understanding of the profound implications for their lives. Relatives and or health/social care professionals generally advise or recommend accepting provision of the equipment. In some cases, older people are given to understand that if they decline such provision in their home, then the only alternative is to be put into a care home. Thus ethical questions raised by the delivery of technology-based care need to be acknowledged and debated, but the pressured world of policy making and service delivery in health and social care for older people is one in which there is unlikely to be time adequately to reflect on them.

4.7   Research on Assistive Technologies

We now turn to ethical issues that arise more specifically from research around the use of assistive technologies with older people. We start by noting the inconsistencies of approach in ethical considerations around assistive technology research in UK universities before discussing research in the human–computer interaction (HCI) design field in the higher education and commercial sectors and the telecare delivery sector.

There continues to be a growth of interest and attention to research ethics internationally. In the UK, this came about in large measure as a result of the wider impact of the Alder Hey inquiry that reported in 1999. The findings revealed by the inquiry were that hospitals within the National Health Service (NHS) were retaining patients' organs without family consent. The inquiry was sparked by the death of 11-month-old Samantha Rickard, who died in 1992 while undergoing open-heart surgery at Bristol Royal Infirmary. Questions about the quality of pediatric cardiac surgery at Bristol led to a public inquiry. The inquiry learned about the large number of hearts held at the Alder Hey Children's Hospital in Liverpool. An investigation was opened in December 1999, and in January 2001, the official Alder Hey report (also known as the Redfern report) was published. It had been found that the unethical and illegal stripping of every organ from every child who had had a postmortem had been ordered by pathologist Dick van Velzen, while he was in post at the hospital. It was also found that over 104,000 organs, body parts and entire bodies of fetuses, and still-born babies were stored in 210 NHS facilities in addition to 480,600 samples of tissue taken from dead patients (BBC News, 2001).

Although this inquiry was concerned with *clinical* research, its impact and ramifications have since been felt more generally across the social sciences. In 2010, the UK Economic and Social Research Council (ESRC) produced a second ethics framework that must be followed by researchers bidding for research funds (ESRC, 2010). This framework includes stipulations such as:

- Research staff and participants must normally be informed fully about the purpose, methods, and intended possible uses of the research; what their participation in the research entails; and what risks, if any, are involved.
- Research participants must take part voluntarily, free from any coercion.
- Harm to research participants must be avoided in all instances.

The importance given by research councils who provide funding is clear thus, for example, the ESRC (2010) notes:

> Breaches of good ethics practice … could result in the immediate suspension of the individual project and other projects based at or under the co-ordination of the contracting institution, and a halt to the consideration of further applications from that institution (p. 4).

Other reasons for the heightened interest amongst social scientists of ethical concerns, particularly regarding research involving older people, may include fear of litigation in the event of negative consequences, growing concern about the scale of care provision required for older people, and the associated drive to implement telecare technologies as a response.

In line with this growth of interest and attention to research ethics internationally (UNESCO-CEPES, 2004) and in the wake of the Research Assessment Exercise (RAE) of 2005 in the UK, Higher Education institutions established Research Ethics Committees (RECs). All proposals for research involving people now have to be approved by the REC of the researcher's institution prior to data collection and/or fieldwork commencing. However, as will be seen, the many ethical frameworks that have evolved in higher education institutions are diverse in the approach they take and in the underpinning principles they reflect.

4.8   Older People's Participation in Research

The ethos regarding the participation of older people in research has changed over the past decade from one where they are regarded primarily as *subjects*, required to provide data for research, to a far more inclusive approach where they are seen as research *participants*. This change has brought with it the growing recognition among some of the research community that the needs and characteristics of older participants in research are deserving of respect and due consideration. It might reasonably be expected that the increased emphasis given over the past decade to the formulation of ethical frameworks and procedures would reflect this concern to safeguard and promote the physical and psychological well-being and dignity of these participants. Recent research (Sus-IT Project, 2010) into the ethical frameworks of a sample of eight UK higher education institutions sheds light on this. It should be emphasized that the documentation scrutinized may not have been the primary ethical policy/procedure documentation available from these institutions, but nevertheless, it was the documentation that participants were aware of and considered most relevant to their work. The documentation was examined for text on four key considerations each of which was considered essential for the ethical engagement of older people in research and informed by long experience of several of the investigators, established expertise of colleagues at the University of Dundee and by key literature (ESRC, 2010; The Belmont Report, 1979; The British Psychological Society, 2009; The Nuremberg Code, 1949). These four principles were:

1. The risk of harm: significant psychological or emotional distress to participants
2. Maximizing benefit (principle of beneficence)
3. Principle of respect for persons, that is, acknowledgement of the dignity of individuals
4. Special consideration of older people in research practices, procedures, and methods

How were these principles used in conducting research in the UK universities under consideration? Drawing on the first category, the attempt to minimize risk to research participants, it appears to be standard practice for participants to be risk assessed in terms of their physical and mental health prior to participating in any research study. The ethical procedures did in most cases explicitly state that any risks that may arise during the study must be explained to participants prior to the study. The ethical framework documentation varied, with only two institutions giving brief instructions regarding how to complete an ethical research protocol. Some ethical framework documentation also mentions the risk that the researcher can pose to the participants. The framework supplied by one university specifies that it is necessary for the researcher to undergo Criminal Records Bureau checks and reference checks prior to consent being given for them to conduct research. The second key consideration, the principle of beneficence, was not mentioned in five of the eight sets of ethical framework documentation. The third key consideration, respect and dignity, did not feature in four out of the eight ethical frameworks examined. The fourth key consideration, explicit consideration of the needs and characteristics of older people, was found in only three of the eight sets of ethical documentation received. One institution clearly states that people over 65 years of age are, by definition, *vulnerable*, but what is meant by vulnerable groups can vary from university to university, and it is not always apparent whether or not older people are automatically considered to be in this category. This is relevant because ethical procedures for dealing with vulnerable people are more demanding. Thus for all the extra layers of ethical safeguards imported into the research process in recent years, there remain significant variations in how research with older people is scrutinized in its ethical dimension.

4.9 Research in the Human–Computer Interaction Design Field: Higher Education and Commercial Sectors

The significant changes in the social, legal, demographic, and economic landscape over the past 15 years present considerable opportunities for the human–computer interaction (HCI) design community to better support people who previously were underrepresented in and consequently whose needs were insufficiently considered in technology design (Newell & Gregor, 2002). There are, however, specific issues around the participation of older people in research, particularly when technology is also involved. These range from effective choice and application of research methods to dealing with discovery of potentially sensitive data and situations involving participants. HCI research rarely reflects demographic reality. Twenty percent of the population in the developed world is over age 60, yet most HCI research is focused on younger people, often university or college students. Rather than representing the population as it actually is, much experimental HCI research is skewed heavily toward the characteristics (and attitudes) of the highly educated young (Dickinson & Gregor, 2006). Equally, not all HCI methods are suitable for use in contexts involving older people as participants, given that older people have an extremely wide range of characteristics and impairments compared to other age groups of participants. Older people are a heterogeneous group. They vary widely in their ages and lifestyles and in their levels of education, independence, and income. They also vary in their range of physical and cognitive abilities (skills). Conducting research with older people gives rise to many challenges. For instance, challenges may arise as a result of individuals' visual impairment, auditory impairment, cognitive changes, and mobility difficulties. Indeed, individuals may be affected by one problem or a combination of several which may increase over time. A user-centered design approach may recognize diversity in characteristics between user groups, but may be less focused on identifying diversity within groups, and particularly so for older people, diversity within an individual's capabilities over the short and long term. The effects of aging will be manifest at different rates relative to one another for each individual. This pattern of capabilities varies widely between individuals and, as people grow older, the variability increases (Myatt, Essa, & Rogers, 2000). In addition, any given individual's capabilities vary in the short term due to a variety of causes including illness, blood sugar levels, and just plain tiredness.

Aging is associated with specific changes in characteristics such as visual and auditory perception, fine motor control and some aspects of memory and cognition (see Hawthorn, 2000 for an overview). While many of these changes are only apparent in psychological tests, some can influence participants' ability to read or hear experimental instructions, use a mouse, or remember steps through an interface. The cultural and experiential gap can be particularly pronounced when involving older people in the development of new technology (Eisma et al., 2003; Malik, Alistair, & Edwards, 2008). Conversely, superior social skills can mean that older participants may be more likely to involve the facilitator in the task (Dickinson & Gregor, 2006). This collection of phenomena presents a fundamental problem for the designers of interfaces to computing systems, whether they be generic systems for use by all ages, or specific systems to compensate for loss of function (Zajicek, 2004), but also has more general implications for researchers involving older people in their research. Age-related capability change has implications for research method selection and design. A lack of familiarity among researchers and designers working with older people can often mean there is a lack of sensitivity in the way in which research with older people is conducted. Although there are data to show that older people can be successfully recruited into beneficial health promotion programs, it is often challenging. In planning health promotion studies, investigators need to be aware of the numbers of older people they may need to screen and different strategies for increasing recruitment success.

There are some key characteristics of older people which need to be reflected in the conduct of research in which their participation is sought, and hence user research methods involving older people should be carefully chosen so as to ensure participants are treated in an ethically sound manner while also maximizing the quality and fidelity of data gathered. Dickinson, Arnott, and Prior (2007) provide some valuable advice, based on personal experience, of conducting HCI research with older people in an ethically and experimentally sound manner, covering selection of appropriate research methods, recruitment, research location, and management of participants before, during, and after the research activity. It is also important to consider user involvement. Lack of user involvement is one of the factors that have been found to lead to abandonment of traditional assistive technologies (Damodaran, Olphert, & Hardill, 2010). An important component of older people's participation and engagement in the design process is decision-making, and its implementation requires the adoption of a participatory approach to socio-technical design. Therefore, older people who are the main intended beneficiaries of specific digital assistive technologies should be able to participate in decision-making not only concerning the technical aspects and system features but also in relation to the policies that affect the delivery and availability of the digital assistive technology (Olphert, Damodaran, Balatsoukas, & Parkinson, 2009). Research, including innovative work on the Sus-IT project has demonstrated the enthusiasm and ability of some older people to participate in codesign activities and to shape design and policy decisions.

Related to the issue of dynamic diversity and age-related capability change, and the implications on research methods used, is the ethical issue of how this can be accommodated in information and communication technology (ICT) and other product design in a way that allows objective research involving older people. Pairing individuals who have accessibility needs with the assistive technology most appropriate to these needs is a particular issue. Aside from any economic difficulties in procuring and using a particular assistive technologies, there is a fundamental issue of awareness. It is assumed that an ICT user with a severe, congenital disability is likely to be fully aware of the assistive technologies they need and has had appropriate support in procuring, installing, and learning to use this technology. However, the obscurity of assistive technologies, and an unwillingness to recognize and address an accessibility need, mean this is less likely to be the case for someone whose visual acuity, dexterity, and short-term memory has gradually declined over many years. The visibility issue is a paradox, given that accessibility features of an ICT can benefit more than people who might be covered by conventional definitions of *disabled* (Forrester Research & Microsoft Corporation, 2003). Dynamic diversity, the unexpected variations over time in an individual's visual, motor, and cognitive capability (Gregor, Newell, & Zajicek, 2002), has been recognized as a key complication in supporting more efficient HCI, and this is most acutely present when considering the unpredictable impact of the aging process on capability. One approach to accommodating the dynamic nature of human capability and the affects it can have in performance with respect to ICT use is to combine user profiling with adaptive interfaces (Gajos, Wobbrock, & Weld, 2007; Sloan, Atkinson, Machin, & Li, 2010). Monitoring user activity and making minor adaptations to system behavior, for example, by giving added prominence to recently or regularly used documents or applications, is a common approach to improving usability, and extending this approach has clear potential for enhancing accessibility.

The ethical challenge to supporting inclusion, through user profiling and adaptation, centers around the capture, storage, and analysis of performance-related data. In a system where regular capability measurements are taken while an individual is using the system, either through automated or semiautomated means, and minor adjustments are made (see Sloan et al., 2010), the altruistic motivation is to allow an existing ICT user to continue to use their ICT independently and successfully and to minimize the chance of age-related capability decline leading to technology abandonment. The technical quality of such a system is dependent on the accuracy of capability measurement made and the success of the reasoning process undertaken to apply an appropriate solution to accommodate an identified capability change; in other words, the system should correctly identify the capability change experienced and apply the best possible solution to ameliorate any negative effects of such change. The result should be a series of fine-grained adjustments that are effectively imperceptible to the user, allowing them to continue to interact with their ICT without significant loss in productivity or enjoyment. An example of such fine-grained change is the system identifying that the user is now having difficulty reading small text sizes and accommodating such difficulties by enlarging the text across the system. Gathering and storing performance-related information in this way has implications, however, particularly given that the data gathered may illustrate an individual's changing capabilities over time—most likely to be a reduction in capability. Further, there is a possibility that regularly sampled capability data may point to an underlying medical condition. To what extent should a system designed to optimize a user's interaction with an ICT deal with data of medical significance? How can a system minimize the negative implications on personal sense of well-being of an individual reviewing data showing their personal capability decline over time? In a study by Sloan et al. (2010), older participants did not raise significant concerns regarding the implications of the data being gathered by the profiling and adaptation system. But longer-term exploration of the impact on personal sense of well-being is an essential aspect of the evaluation of any such system. Possible approaches to addressing the ethical implications of these questions might include:

1. Seeking informed consent after their first use of a capability monitoring and adaptation system
2. Seeking approval for every adjustment made based on a capability measurement, although depending on the frequency of interruptions, this approach may have negative implications on productivity
3. Secure and anonymous storage of profile information
4. Presentation of profile information in a way that minimizes misinterpretation by the individual concerned or any other approved viewer
5. Limiting capability measurements to those that directly map to an accessibility adjustment supported by the host system

The first step in this process is to identify the potential implications of a system aimed at sustaining independent access by people with accessibility needs and consider carefully the data to be collected, in terms of nature, analysis, and storage.

There is thus a need for a paradigm shift from current ICT design approaches that focus on technical aspects of system design (e.g., based upon a waterfall model) and that would engage older people at specific stages of the design process, such as usability evaluation or task-requirements analysis, to a socio-technical approach that values human participation throughout the design process and elaborates on the dynamics of older people's needs across time. Socio-technical theory approaches technology as a complex system, where technical elements interact with social and organizational aspects of the system. Thus, the social or organizational context can influence the development and implementation of technology. Older people should be an integral component of the socio-technical system, and their role within it should be extended from merely participating in the evaluation of assistive technologies to decision-making, learning, and knowledge sharing as well as communicating their beliefs, aspirations, and expectations about these technologies to other groups of stakeholders (Olphert et al., 2009).

4.10   Assistive Technologies and Quality of Life

As noted at the outset of this chapter, there is an assumption in the UK that increased deployment of assistive technologies with older people can positively impact on QoL. The term is employed explicitly, for example, in government policy documents and in research papers on telecare provision. The final section of this chapter examines this assumption. The idea of QoL as an ethical issue emerged with Aristotle in his Nicomachean Ethics, which argued that the study of ethics was essentially about the (in Aristotle's case, virtuous) pursuit of human well-being and in finding ways to improve people's lives. Thus, the policy documents that espouse a QoL connection are broaching a deeply ethical question. Thus broached, we now ask in what way, if at all and if measurable, might assistive technologies actually impact on QoL? Unless we can be clear about the criteria that underpin QoL measurement itself, and how assistive technologies might connect to this, the claim that the use of assistive technologies can enhance it is a bold one, as although the term *quality of life* pervades discussion of health and social policy, there is little consensus on what the term means, how best to measure it, and how best to increase it at an individual and national level. Here, we explore definitional and measurement challenges associated with QoL, and reflect on research about the role of assistive technologies in QoL in old age.

Research on QoL began in the first half of the twentieth century and was aimed at measuring population well-being, with measures such as gross domestic product per capita considered to be aspects of life quality. However in the 1960s, there was a shift to much broader indicators and the inclusion of QoL as a characteristic of persons as well as national prosperity (Rapley, 2003). In tandem with this, there was greater acceptance of the role of government in actively shaping societal structures for the greater good. A second phenomenon occurring in the 1960s was an increase in the number of expensive medical treatments for a variety of diseases and chronic conditions. Many of these treatments could not cure. Questions began to be asked not only about how these interventions could be evaluated but how cost-effective they might be. The late 1960s saw the introduction of the *quality-adjusted life year* movement in medicine, and while it was accepted that many treatments could not cure, they could perhaps be shown to increase patients' QoL. The findings from early studies were, however, paradoxical. Study after study found that people in poor living conditions reported high life satisfaction and people with serious disabilities reported a high QoL. Moreover, increasing wealth across the world was not increasing perceptions of increased QoL, except amongst the very poor. It was clear that social comparisons and aspirations were influencing perceptions of QoL. The many instances of the satisfaction paradox and the disability paradox led to programs of research to delineate the domains of QoL. One of the biggest and most expensive of these was the World Health Organization's (WHO) Quality of Life Project that was initiated in 1991. This project identified the following broad domains: physical health, psychological health, social relationships, and environment (WHO, 1993). The scales developed within the WHO project were designed to consider the context of the culture and value systems of those who were rating, as well as personal goals, standards, and concerns. Other research projects have found similar domains, for example, Nazroo, Bajekal, Blane, Grewal, and Lewis (2003) revealed six factors influencing QoL at older ages: having a role, support networks, income and wealth, health, having time, and independence. These attempts to determine the domains of QoL result in, or are a consequence of, a confounding of predictors and domains. Researchers often determine the domains from findings about predictors; only a few have started with a theoretical perspective. Blane, Wiggins, Higgs, and Hyde (2002) are among the few that have; their QoL measure was based on the theory of needs satisfaction and included the domains of control, autonomy, self-realization, and pleasure. Broadening the scope of the term QoL to include a number of domains has left us with a slippery, complex concept which is now used to describe everything (Rapley, 2003), but which has no agreed definition. With no agreed definition, there can be no agreed method of measurement.

Given the difficulties in defining and measuring QoL, it is not surprising that it has been challenging to find research evidence that assistive technologies enhance QoL. Compounding the definitional and measurement problems is the wide range of assistive technology available and the poor quality of many of the studies that have been conducted. For example, in a recent review of the literature on environmental control systems and smart home technologies, Brandt, Samuelsson, Toytari, and Salminen (2010) found 1,739 studies, but only five effect studies and six descriptive studies met their criteria for selection. Of these 11 studies, only one examined QoL, and it was not a study of older people. Dickinson and Gregor (2006) published a critique of the value of personal computers to older people's QoL and well-being in which they concluded that computer use appears not to lead to improved well-being. As was the case with the review by Brandt et al. (2010), they found few studies to review, and those that were available had many weaknesses. Dickinson and Gregor (2006) also examined how the authors of secondary papers cited the findings and the extent to which they offered a sound critique. What was particularly interesting was that many authors citing the original studies suggested that computers had indeed made a positive impact, when in fact the original studies could not or did not support such claims. A literature review by Ridley and Young (2005) examined the effectiveness of e-health implementation in the care of elderly people. The review identified 647 primary articles, and 66 met the review criteria. Most of the studies focused on teleconferencing for particular health conditions. Ridley and Young concluded that there was sufficient evidence that these technologies had positive outcomes. QoL was not actually measured in most of these studies. Instead, the impact on QoL could only be inferred from outcomes such as patient control, psychological support, greater functional independence, reduced accident and emergency admissions, reduced levels of depression, and reduced need for patient travel. The small numbers in most of the studies reviewed, however, limited the impact of the studies. Studies on the impact of telemedicine on the health and QoL of older adults were reviewed by Jennett et al. (2003), who concluded that telemedicine can improve the QoL and health of older people. However, a large number of telemedicine initiatives have failed because they have been set up in isolation and without thought as to their cost-effectiveness (Macduff, West, & Harvey, 2001), let alone their impact on QoL.

4.11 Why Assistive Technologies Are Unlikely to Influence Quality of Life Ratings

There are a number of ways of looking at the research findings on the impact of assistive technologies (ATs) on QoL ratings. It could be, particularly in the face of so little evidence, that ATs correlate positively with QoL ratings:

- ATs might have specific effects but not general effects.
- ATs do not influence QoL.
- QoL is a meaningless concept, and hence, it is nonsensical to assess the impact of ATs on QoL.

Each of these hypotheses will now be considered in turn. It could be that ATs have specific effects, and indeed, there is some evidence to support this view. Alternatively, it could be that the current QoL measures are simply too remote, blunt, and multifactorial to be affected by ATs. It could also be that a high QoL brings about an interest in and use of ATs. This is certainly likely to be the case with personal computer use. In all correlational research, we must be vigilant not to read the correlation in only one direction. There is also a need to explore ways in which a technological divide may exist between older people who are familiar and comfortable with technology and those who are reluctant to embrace it or find it harder to access. Technology might lower QoL for the latter. Returning to the previously discussed telecare development agenda in the UK, we can note the presumption in the policy that telecare will enhance older people's lives. But exploratory research across different geographical areas of telecare implementation suggests a varied picture of the willingness by older people to engage with technology, a key element here being prior exposure (e.g., in the workplace) (Eccles, 2010). It should also be noted that health, wealth, and social relations consistently emerge as factors that predict ratings on QoL scales. Young people without chronic health problems have been found to be somewhat less likely to rate health as a prime determinant of QoL. For older people, however, poor health accounts for a high proportion of variance. Given that age is the main risk factor for almost all illness, aging without experiencing poor health is rare (Manton, 1989; Wood & Bain, 2001). Thus, while the exponential rise in disease incidence indicates a greater need for ATs in old age, the impact of poor health may be so great, and account for such a high proportion of the variance in QoL ratings, that it becomes almost impossible (statistically) for ATs to make an impact on QoL ratings.

In addition, as people age, social relations (particularly those with family members) become very important. As people become frail and in need of family support, social relations may suffer or become problematic. It is now fairly well established that negative social relations have a greater impact on psychological well-being than positive relations. Older people want to be independent and may resent offers of help from adult children. Older people are frequently embarrassed at having to ask for help. Another interesting possibility is that happiness and/or QoL is a dispositional characteristic (Costa & McCrae, 1980; Costa, McCrae, & Zonderman,

1987) that is not only stable across the life span but may be biologically determined (Diener, 2000). Returning to the ethical issue—if QoL has no agreed definition, and there are serious measurement issues, is it sensible, or indeed ethical, to view improved QoL as an outcome of policy interventions involving ATs? As Rapley (2003) notes, to use the QoL construct to gauge the success of ATs when QoL is viewed as an individualized aspect of the modern psyche is paradoxical. If QoL is individualized how can a case be made to assess it in the same way for everyone? We suggest a more nuanced enquiry might be: What are the predictors of QoL, and how might ATs connect to these?

4.12 Conclusion

This chapter has raised a number of issues around ethical research and ethical practice in the design and application of gerontechnology. As in any field where there is rapid technological development with an application to human services, there is the potential for ethical issues to arise that could not easily have been anticipated and for the limitations of existing approaches to ethical codes to be exposed. This in itself is not problematic. The issue is how this can be addressed, particularly when technological advances continue apace, not least through the agendas pursued by equipment manufacturers, but where ethical enquiry remains a less mainstream pursuit. We note evidence of the profound impact that gerontechnology can have in improving some people's lives. We note also the concern that where its uptake is stymied by poor awareness, or adherence to long established ways of working in the field of social care, this in itself may pose an ethical problem, as it means that older people might not be gaining these advantages. We argue, furthermore, that there is a risk of creating a false dichotomy between technology-based care and human care, as if one is inherently superior to the other. Context is important here, and independence is highly prized and may be impinged upon by reliance on carers, the inconvenience of care organizations' schedules, or the inability to age in situ. Assistive technologies clearly confer benefit here. This said, we note a number of areas where, drawing on our experience primarily of the situation in the UK, we suggest that there are deficits and potential problems in ethical engagement. These are fourfold. First, there is a need for more phenomenological research into how people experience the use of technology. We note the proliferation of ethics committees involved in the research process but a concomitant weakness with their understanding of some aspects of qualitative research, and we note also the apparent risk-averse nature of ethical approval in some universities, where older people may automatically be classed as vulnerable participants. Second, we have raised the concern that technological solutions represent too much of an ideal fix for the twin policy challenges of demographic change and rising costs of elder care. A powerful discourse has developed here, evidenced particularly in the literature of leading AT corporations and in policy forums of social care budget holders, where the potential offered by these technologies becomes subsumed by assumptions of their necessity.

We believe, drawing on the UK context, policy on the introduction of care technologies to be ill-served by the performance indicator regimes put in place by government which may encourage an emphasis on rapid expansion of deployment at the expense of suitability and context. An additional factor here is the often rapid change in the circumstances of older people and the need for personnel to oversee regular review of the suitability of technologies that have been installed. Third, we believe the ethical frameworks employed by agencies that carry out assessments for older people to be limited. There is a balance to be struck here between ethical codes that are overly complex but are nonetheless fit for purpose. That notwithstanding, the biomedical parameters predominantly in use lack sophistication in their application to the complexities of long-term care conditions and additional ethical enquiry based on, for example, more relational approaches would be worth exploring. The ethical issues here are compounded by the increasing moves toward interprofessional working, where similar tasks may be carried out by one of a number of agencies, each with their own particular codes of professional inquiry. Fourth, we note the explicit connection in government policy documents between the deployment of ATs and enhanced QoL. There are important issues raised here, not least equity of access to technology and assumptions that technology use will be equally well embraced, or readily useable, by older people. Equally, measuring QoL indicators and connecting these to specific technological interventions is fraught with methodological complexities. We would thus urge caution in the way that this term is employed. There is certainly evidence that some people in some circumstances enjoy better lives through the use of assistive technologies. To make broader claims on this issue at this stage risks the ethical deficit of foreclosing further critical enquiry and reflection.

References

Audit Commission. (2004). *Implementing telecare*. London: Audit Commission.
Banks, S., & Gallagher, A. (2009). *Ethics in professional life: Virtues for health and social care*. Basingstoke: Palgrave Macmillan.
Barnes, M. (2006). *Caring and social justice*. Basingstoke: Palgrave MacMillan.
BBC News. (2001). *The Alder Hey report*. Retrieved from http://news.bbc.co.uk/2/hi/health/1144774.stm
Beauchamp, L., & Childress, A. F. (2001). *Principles of biomedical ethics* (5th ed.). Oxford: Oxford University Press.
Blane, D., Wiggins, R., Higgs, P. & Hyde, M. (2002). Inequalities in quality of life in early old age. *Research Findings from the Growing Older Programme, 9*.
Blaschke, C., Freddolino, P., & Mullen, E. (2009). Ageing and technology: A review of the research literature. *British Journal of Social Work, 39*, 641–656.
Bouma, H. (2010). Professional ethics in gerontechnology: A pragmatic approach. *Gerontechnology, 9*(4), 429–432.
Brandt, A., Samuelsson, K., Toytari, O., & Salminen, A-L. (2010) Activity and participation, quality of life and user satisfaction outcomes of environmental control systems and smart home technology: A systematic review. *Disability and Rehabilitation: Assistive Technology*, Early Online, 1–18.

Cooper, A. (2012). Complexity, identity, failure: Contemporary challenges to working together. In J. Forbes & C. Watson (Eds.), *The transformation of children's services*. New York: Routledge.

Costa, P. T., & McCrae, R. R. (1980). Influence of extraversion and introversion on subjective well-being: Happy and unhappy people. *Journal of Personality and Social Psychology, 38*, 668–678.

Costa, P. T., McCrae, R. R., & Zonderman, A. B. (1987). Environmental and dispositional influences on well-being: Longitudinal follow-up of an American national sample. *British Journal of Psychology, 78*, 299–306.

Damodaran, L., Olphert, C. W., & Hardill, I. (2010). Some ethical considerations about informed consent by older people to assisted living technologies and the participation of older people in ICT research. *Gerontechnology, 9*(2), 85–86.

Dickinson, A., Arnott, J., & Prior, S. (2007). Methods for human-computer interaction research with older people. *Behaviour & Information Technology, 26*(4), 343–352.

Dickinson, A., & Gregor, P. (2006). Computer use has no demonstrated impact on the well-being of older adults. *International Journal of Human Computer Studies, 64*, 744–753.

Diener, E. (2000). Subjective well-being: The science of happiness and a proposal for a national index. In M. E. P. Seligman & M. Csikszentmihalyi (Eds), Special issue on happiness, excellence, and optimal human functioning. *American Psychologist, 55*, 34–43.

Driver, J. (2007). *Ethics: The fundamentals*. Oxford: Blackwell.

Eccles, A. (2008). Singe shared assessment: The limits to 'quick fix' implementation. *Journal of Integrated Care, 16*(1), 22–30.

Eccles, A. (2010, April). *Ethical considerations around the implementation of Telecare technologies*. Paper presented at SICSA Socio-Technical Systems Conference, University of Glasgow, Glasgow.

Eisma, R., Dickinson, A., Goodman, J., Mival, O., Syme, A., & Tiwari, L. (2003). Mutual inspiration in the development of new technology for older people. *Include, March*, 7:252–7:259.

Equality and Human Rights Commission. (2011). *Close to home: An enquiry into older people and human rights in home care*. London: EHCR.

ESRC. (2010). *Framework for Research Ethics (FRE)*. Retrieved February 9, 2011 from http://www.esrc.ac.uk/_images/Framework_for_Research_Ethics_tcm8-4586.pdf

European Union Public Health Information System. (2009). *Population projections*. Retrieved February 5, 2010 from http://www.euphix.org/object_document/o5116n27112.html

Finch, T., Mort, M., Mair, F. S., & May, C. (2008). Future patients? Telehealthcare, roles and responsibilities. *Health & Social Care in the Community, 16*(1), 86–95.

Forrester Research & Microsoft Corporation. (2003). *New research study shows 57 percent of adult computer users can benefit from accessible technology*. Retrieved February 1, 2010 from http://www.microsoft.com/presspass/press/2004/feb04/02-02AdultUserBenefitsPR.mspx

Frisby, B. (2000). *ASTRID: A guide to using technology in dementia care*. London: Hawker.

Gajos, K. Z., Wobbrock, J. O., & Weld, D. S. (2007). Automatically generating user interfaces adapted to users' motor and vision capabilities. In *UIST'07: Proceedings of the 20th Annual ACM Symposium on User Interface Software and Technology* (pp. 231–240). New York: ACM Press.

Gregor, P., Newell, A. F., & Zajicek, M. (2002). Designing for dynamic diversity: interfaces for older people. In *ASSETS'02: Proceedings of the Fifth International ACM Conference on Assistive Technologies* (pp. 151–156). New York, NY: ACM Press.

Hanson, J., Osipovic, D., & Percival, J. (2009). Making sense of sensors. In B. Loader, M. Hardey, & L. Keeble (Eds.), *Digital welfare for the third age* (pp. 91–111). London: Routledge.

Hawthorn, D. (2000). Possible implications of aging for interface designers. *Interacting with Computers, 12*, 507–528.

Irvine, R., Kerridge, I., McPhee, J., & Freeman, S. (2002). Interprofessionalism and ethics: Consensus or clash of cultures? *Journal of Interprofessional Care, 16*(3), 199–210.

Jennett, P. A., Affleck Hall, L., Hailey, D., Ohinmaa, A., Anderson, C., Thomas, R., et al. (2003). The socio-economic impact of telehealth: A systematic review. *Journal of Telemedicine and Telecare, 9*(6), 311–320.

Loader, B. (2005). Necessary virtues: The legitimate place of the state in the production of security. Retrieved from http://www.libertysecurity.org/article232.html

Loader, B., Hardey, M., & Keeble, L. (Eds.). (2009). *Digital welfare for the third age*. London: Routledge.

Lowe, C. (2009). Beyond telecare: The future of independent living. *Journal of Assistive Technologies, 3*(1), 21–23.

Macduff, C., West, B., & Harvey, S. (2001). Telemedicine in rural care. Part 2: Assessing the wider issues. *Nursing Standard, 15*(33), 33–38.

Malik, S. A., Alistair, D., & Edwards, N. (2008). Retrieved January 7, 2011 from http://wwwedc. eng.cam.ac.uk/~jag76/hci_workshop08/malik.pdf

Manton, K. (1989). Epidemiological, demographic and social correlates of disability among the elderly. *The Millbank Quarterly, 67*(Siupp.2 PI), 13–58.

McCreadie, C., & Tinker, A. (2005). The acceptability of assistive technology to older people. *Ageing and Society, 25*, 91–110.

Meagher, G., & Parton, N. (2004). Modernising social work and the ethics of care. *Social Work and Society, 2*(1), 10–27.

Mort, M., Roberts, C., & Milligan, C. (2009). Ageing, technology and the home: A critical project. *European Journal of Disability Research, 3*, 85–89.

Myatt, E. D., Essa, I., & Rogers, W. (2000). *Increasing the opportunities for ageing in place, in: Proceedings of the ACM conference on universal usability* (pp. 39–44). New York/Washington, DC: ACM Press.

Nazroo, J., Bajekal, M., Blane, Grewal, I., & Lewis, J. (2003). *Ethnic inequalities in quality of life at older ages: Subjective and objective components*. Research Findings: 11. From the Growing Older Programme, January. Retrieved from http://www.shef.ac.uk/uni/projects/gop/index/htm

Newell, A. F., & Gregor, P. (2002). Design for older and disabled people—where do we go from here? *Universal Access in the Information Society, 2*, 3–7. Retrieved December 14, 2010 from http://www.springerlink.com/content/8c279eqq580uc9vf/

Olphert, C. W., Damodaran, L., Balatsoukas, P., & Parkinson, C. (2009). Process requirements for building sustainable digital assistive technology for older people. *Journal of Assistive Technology, 3*(3), 4–13.

Phillips, J. (2007). *Care*. Oxford: Polity.

Pols, J., & Moser, I. (2009). Cold technologies versus warm care? On affective and social relations with and through care technologies. *European Journal of Disability Research, 3*, 159–178.

Rapley, M. (2003). *Quality of life research: A critical introduction*. London: Sage.

Ridley, G., & Young, J. (2005). Outcomes of international e-health implementations. In *Aged Care, Proceedings of NACIC'05 National Aged Care Informatics Conference*, Hobart, Australia.

Scottish Government. (2008). *Telecare in Scotland: Benchmarking the present, embracing the future*. Edinburgh: Crown Copyright.

Scottish Government. (2009). *Evaluation of the telecare development programme: Final report*. Edinburgh: Crown Copyright.

Sloan, D., Atkinson, M., Machin, C., & Li, Y. (2010, April). The potential of adaptive interfaces as an accessibility aid for older web users. *Accepted for presentation at: W4A 2010: The 7th International Cross-Disciplinary Conference on Web Accessibility*, Raleigh, NC, USA.

Sus-ITproject. (2010). Retrieved from http://www.newdynamics.group.shef.ac.uk/sus-it.html. Retrieved December 14, 2010 from http://www.sus-it.lboro.ac.uk/

Tao, J., & Drover, G. (1997). Chinese and western notions of need. *Critical Social Policy, 17*, 5–25.

Telecare Services Association. (2010). *The true value of telecare services*. Accessed November 29, 2011 from http://www.telecare.org.uk/information/45779/48672/48675/the_true_value_of_ telecare_services/

The Belmont Report. (1979). *Ethical principles and guidelines for the protection of human subjects of research*. Accessed February 11, 2011 from http://ohsr.od.nih.gov/guidelines/belmont. html#goc1

The British Psychological Society. (2009). *Code of ethics and conduct*. Retrieved from http:// www.bps.org.uk/document-download-area/document-download$.cfm?file_uuid=E6917759- 9799-434A-F313-9C35698E1864&ext=pdf

The Nuremberg Code. (1949). Retrieved February 11, 2011 from http://ohsr.od.nih.gov/guidelines/nuremberg.html

Tronto, J. (1994). *Moral boundaries: A political argument for an ethic of care*. London: Routeledge.

Tunstall, R. (2009). Assessing local authority and neighborhood variations in the likely impact of the recession on renters and landlords. http://www.york.ac.uk/inst/chp/HAS/Paper%209%20%20LA%20&%20neighbourhood%20variations%20.pdf

Twigg, J. (1999). The spatial ordering of care: Public and private in bathing support at home. *Sociology of Health & Illness, 21*(4), 381–400.

UNESCO-CEPES. (2004). *Evaluation of the UNESCO-CEPES*. Retrieved from http://unesdoc.unesco.org/images/0014/001449/144949e.pdf

Wilmot, S. (1997). *The ethics of community care*. London: Cassell.

Wood, R., & Bain, M. (2001). *The health and well-being of older people in Scotland: Insights from national data*. Edinburgh: Information and Statistics Division, Common Services Agency for NHS Scotland.

World Health Organization Quality of Life Group. (1993). *Measuring quality of life: The development of the World Health Organization Quality Of Life Instrument (WHOQOL)*. Geneva: World Health Organization.

Zajicek, M. (2004). Successful and available: Interface design exemplars for older users. *Interacting with Computers, 16*(3), 411–430.

Chapter 5
Expanding the Technology Safety Envelope for Older Adults to Include Disaster Resilience

Maggie Gibson, Gloria Gutman, Sandra Hirst, Kelly Fitzgerald, Rory Fisher, and Robert Roush

5.1 Introduction

The concept of resilience is central to understanding how technology might have a role to play in reducing the disproportionate vulnerability of older adults in natural and human-made disasters. Resilience has been defined in various ways by different theorists and researchers, but the common thread is the idea of adaptive capacity

M. Gibson, Ph.D., C.Psych (✉)
Veterans Care Program, Parkwood Hospital, St. Joseph's Health Care London,
801 Commissioners Road East, London, ON, Canada N6C 5J1
e-mail: maggie.gibson@sjhc.london.on.ca

G. Gutman, Ph.D.
Gerontology Research Centre, Simon Fraser University, Harbour Centre,
#2800-515 West Hastings Street, Vancouver, BC, Canada V6B 5K3
e-mail: gutman@sfu.ca

S. Hirst, R.N., Ph.D., G.N.C.(C)
Brenda Strafford Centre for Excellence in Gerontological Nursing, University of Calgary,
2500 University Dr. NW, Calgary, AB, Canada T2N 1N4
e-mail: shirst@ucalgary.ca

K. Fitzgerald, Ph.D.
Western Kentucky University, Alte Hedingerstrasse 44a, Affoltern am Albis,
8910, Switzerland
e-mail: kellyfitzgerald@hotmail.com

R. Fisher, M.B., F.R.C.P. (C), F.R.C.P. (Ed.)
Division of Geriatric Medicine, Department of Medicine, Sunnybrook Health Science Centre,
University of Toronto, 2075 Bayview Avenue, Toronto, ON, Canada M4N 3M5
e-mail: rory.fisher@sunnybrook.ca

R. Roush, Ed.D., M.P.H.
Department of Medicine, Baylor College of Medicine, Huffington Center on Aging,
One Baylor Plaza, MS230, Houston, TX, USA
e-mail: rroush@bcm.edu

A. Sixsmith and G. Gutman (eds.), *Technologies for Active Aging*,
International Perspectives on Aging, DOI 10.1007/978-1-4419-8348-0_5,
© Springer Science+Business Media New York 2013

and the ability to recover from adversity (Norris, Stevens, Pfefferbaum, Wyche, & Pfefferbaum, 2008). Resilience is not simply a personality style or a characteristic of individuals but a product of the interplay among various determinants of population health: income and social status, social support, education, employment, social and physical environments, health practices and coping skills, developmental factors, biological and genetic endowment, health services, gender, and culture (Public Health Agency of Canada [PHAC], 2003). Disasters are large-scale disturbances or sources of adversity that tax the resilience not only of individuals but of whole communities and broader societies. All members of a population can be at risk depending on the nature of the crisis. As would be expected given the multiple interacting determinants that come into play, the pathways from risk vulnerability to disaster resilience are complex (International Federation of Red Cross and Red Crescent Societies [IFRC], 2004). Similar to other population health challenges, the availability of appropriate resources, effectively implemented, is likely to contribute to more desirable outcomes for individuals and groups who are responding to and attempting to recover from disasters (Lindsay, 2003). Disaster resilience is increasingly in the public eye as the number of catastrophic natural and human-made events continues to rise. This chapter examines the potential for technology to promote disaster resilience among older adults. They are a population subgroup with increased vulnerability in emergencies not because of age per se, but because they are more likely to live with a constellation of risk factors for increased vulnerability, including health problems, dependence on healthcare and social services, lower socioeconomic status, and restricted social networks. In addition, with increasing age, higher proportions of older adults are women, a population subgroup with heightened vulnerability across the life course (Powell, 2009).

Research shows that older adults are given low priority and little attention before, during, and after disasters (HelpAge International, 2000), despite experiencing disproportionate risk for morbidity and mortality. The neglect of older adults' unique concerns is thought to be systematic and discriminatory (IFRC, 2007). The consensus from two international workshops on seniors and emergency preparedness convened by the Public Health Agency of Canada was that there was a critical need to ensure older adults are included in all aspects of emergency management (PHAC, 2008a, 2008b). Similar conclusions have been expressed in publications by AARP (Gibson & Hayunga, 2006), the World Health Organization (WHO) (Hutton, 2008; WHO, 2008), and the IASC (2008). The gaps identified in this literature are far reaching and include lack of consideration of older adults' vulnerabilities such as their ease of access to healthcare and social services, economic subsidies and incentives, and transportation. We argue that there is a role for technology in bridging some of the gaps.

At the same time, it should be noted that there is also insufficient attention given to the contributions that older adults can make to their families and communities in disaster situations. For example, in *Older People in Disasters and Humanitarian Crises: Guidelines for Best Practice*, HelpAge International (2000) notes that "Older people play valuable roles as carers and resource managers, while the knowledge they hold—of traditional survival systems, appropriate technologies, and alternative medicines—can be central to the development of community coping strategies in and after crises" (p. 12). Older adults' experience and their willingness to participate in emergency response and recovery activities represent largely untapped resources in both traditional and more industrialized societies. In industrialized societies in particular, there is potential for the increasingly tech-savvy older generation to contribute to disaster resilience through participation in research and knowledge translation that focuses on the applications of technology to emergency management challenges.

To set the stage for discussion of the role of technology in building and facilitating disaster resilience among older adults, we begin by describing the four phases of the emergency management cycle and the risk factors that contribute to increased vulnerability for older adults. Our analysis is framed within the emergency management cycle to highlight the potential for technological solutions to contribute to the resolution of practical problems that disproportionately impact older members of the population, and thereby to increase the likelihood that they will survive a disaster and recover in its aftermath.

The technologies we consider include tracking and mapping systems, intelligent building systems, medical and assistive devices, communication and notification systems, needs assessment strategies, medical support strategies, security strategies, and reconstruction strategies. The chapter closes with a discussion of factors that may influence the acceptance, uptake, and application of technology in building disaster resilience in aging individuals and the systems that they rely on for support and protection.

5.2 The Emergency Management Cycle

The four phases of the emergency management cycle are prevention/mitigation, preparedness, response, and recovery (WHO, 2008). *Prevention and mitigation* entail proactive measures taken before an emergency situation occurs in the interests of eliminating or reducing the potential impacts and risks. *Preparedness* refers to actions taken to become ready to respond to an emergency situation if and when one occurs. Prevention, mitigation, and preparedness activities target risk reduction (WHO, 2007). *Response* refers to the actions that are taken during and immediately after an emergency situation in order to manage the consequences, and in particular to limit harm and loss, including loss of life. *Recovery* involves efforts to repair or restore conditions to an acceptable level through measures taken after a disaster.

Although it is useful to conceptualize emergency management within these four phases for planning purposes, in actual crises, there is significant overlap. For example, in a large-scale flood, different communities, discrete households within communities, and individuals within those households may be at very different stages of action. Some communities, households, and individuals will be well prepared, have ample resources and strong social supports, and be able to move relatively quickly from response to recovery, while others may be scrambling to prepare as the flood waters rise and/or be lost in bureaucracy when the window of opportunity for restoration support is open. Gaps in emergency management can contribute to a hazardous situation becoming a disaster involving widespread injury, property damage, and death.

5.3   Vulnerability Risk Factors for Older Adults

Internationally, there is mounting evidence that older adults are at increased risk for morbidity and mortality in emergency situations (Barney & Roush, 2009). For example, the WHO (2008) reports that the highest age-specific death rates in Aceh, Indonesia, that resulted from the 2004 Indian Ocean tsunami were among persons aged 60–69 (22.6 %) and 70 and over (28.1 %). The European heat wave of 2003 caused 14,800 deaths in France alone, and it was older adults who had the highest mortality (Kosatsky, 2005). Almost three quarters of the deaths from Hurricane Katrina in the United States in 2005 were in those aged 60 and older (Gibson & Hayunga, 2006). Physical and medical issues that increase the vulnerability of older adults include mobility limitations, changes in thermoregulation ability, and chronic disease. Impaired mobility reduces physical capacity to take evasive and defensive action, such as evacuating a high-rise building using stairs (Shields, Boyce, & McConnell, 2009). Age-related changes in thermoregulation place the older adult at risk of either hyperthermia or hypothermia, although few die from these causes (Goodwin, 2007). In fact, the major causes of death during cold weather are respiratory (e.g., chronic obstructive pulmonary disease, bronchitis) and thrombotic illnesses (e.g., myocardial infarction, stroke). Relatively minor exposure to cold in daily life increases hypertension and hemoconcentration (Donaldson, Robinson, & Allaway, 1997). This may explain why deaths from arterial disease are more prevalent in the northern hemisphere in winter. Impaired cardiovascular reflexes are also implicated in hot weather-related mortality. The risk here however is lowered blood pressure.

Many older adults have chronic diseases such as pulmonary and cardiovascular disease that require ongoing treatment. In 2009, nearly a quarter of Canadian seniors (23 %) indicated that they had some form of heart disease (PHAC, 2010). Coronary artery disease leads to angina and myocardial ischemia, causing decreased physical endurance. Hypertension, cardiac arrhythmias, and congestive heart failure are also common, causing reduced physical activity and fatigue. Cerebrovascular disease can lead to stroke, resulting in residual deficits and weaknesses that may require the use of mobility aids such as canes, walkers, and wheelchairs and/or the assistance of others to maintain independence.

Osteoarthritis, affecting an estimated 85 % of persons aged 75 and older (PHAC, 2010), causes pain, limits mobility, and is a risk factor for falls. Osteoporosis, which leads to loss of bone mass and an increased risk of fractures, is also common in the older population, affecting an estimated 65 % of women and 6 % of men aged 65 and over. Many older adults also have vision and hearing impairments that can compromise their ability to respond to danger.

Dementia and frailty are specific risk factors for heightened vulnerability. Dementia is the name given to a group of progressive neurological diseases that slowly destroy memory and reasoning, erode independence, and eventually take life. Alzheimer's disease is the most well-known and common form of dementia (Alzheimer Society of Canada, 2010). Impaired cognition resulting from dementia reduces decision-making and follow-through capacity in a disaster situation. Frailty is a clinical syndrome separate from the normal aging process in which impairments such as sarcopenia, functional decline, neuroendocrine dysregulation, and immune impairments occur in combination (Abellan Van Kan et al., 2008). The comorbid aspects of dementia and frailty exacerbate each condition, making the provision of both physical and mental health care even more problematic in disaster situations.

Risk factors related to health intersect with situational and socioeconomic factors to increase vulnerability for older adults in emergencies and disasters. Social isolation is one of the most important of the latter (Gibson & Hayunga, 2006; HelpAge International, 2000). Older adults who live alone who have physical, cognitive, or mental health conditions that limit their functional abilities may face insurmountable barriers in meeting practical challenges such as obtaining heat, electricity, potable water, food, and medical supplies and providing pet care in emergency situations. They may be reliant on healthcare and social services that are disrupted by the event. Older immigrants may be at particular risk for experiencing cultural, linguistic, and/or literacy barriers that reduce their access to information and resources. Generally, older adults have fewer financial and social resources to draw on. Moreover, lack of awareness, ageism, and systemic discrimination contribute to the neglect of vulnerable older adults in emergency preparedness initiatives and in community recovery and rehabilitation activities post-disaster (Hutton, 2008).

Older adults who live in long-term care facilities also face distinct challenges during and after disasters. Residents of these facilities have serious physical and/or cognitive impairments, rely on others for assistance with activities of daily living, and require 24 h skilled nursing supervision. One in five Canadians aged 85 years and older currently reside in long-term care facilities (Statistics Canada, 2010). However, it is estimated that by 2041, 120,000 beds will be needed in addition to the current 200,000 long-term care beds across the country. The frailty and medical complexity of facility-based residents are very different from what they were a decade or two ago. Residents are admitted when they are closer to the end of life. They are more functionally dependent and require greater assistance with activities of daily living (Frohlich, De Coster, & Dik, 2002; McGregor et al., 2010; Smith, Tremethick, Johnson, & Gorski, 2009). Many have a diagnosis of dementia.

The effects of Hurricane Katrina demonstrated the vulnerability of long-term care facility residents during and after disasters. Common problems in these facilities included inability to track and monitor all residents; loss of electricity for brief or sustained periods of time; lack of sufficient and appropriate transportation for evacuation of residents; disruption of communication systems with breakdowns in telephone and cellular services; lack of food, water, medications, oxygen equipment, and other medical and general supplies; lack of sufficient numbers of staff; and lack of adequately prepared staff (Deeg, Huizink, Comijs, & Smid, 2005; Dosa, Grossman, Wetle, & Mor, 2007; Laditka, Laditka, Cornman, Davis, & Richter, 2009; Saliba, Buchanan, & Kingston, 2004). Evidence suggests that in the case of Hurricane Katrina and other disasters, long-term care facilities receive less support than needed from emergency response agencies (Brown, Hyer, & Polivka-West, 2007; Dosa et al., 2007; Hyer, Brown, Berman, & Polivka-West, 2006; Laditka et al., 2009).

5.4   Gerontechnology and Emergency Management

Gerontechnology is an interdisciplinary field of research and application involving gerontology-the scientific study of aging, and the development and distribution of technology-based products, environments and services (Fozard, Rietsema, Bouma, & Graafmans, 2000). In general, much of the work on human interaction with machines, devices, and information systems can be usefully conceptualized as an attempt to maximize the degree of fit between the hardware, software, and instructional components of technological systems and the user's sensory, perceptual, cognitive, and psychomotor abilities (Czaja, Sharit, Charness, Fisk, & Rogers, 2001). As a discipline, gerontechnology has acquired depth and breadth over the past two decades (Charness & Jastrzembski, 2009). In its essence, however, it continues to primarily involve the integration of two broad fields—engineering and gerontology—to achieve practical goals such as providing solutions to compensate for deficits in motor functioning, sensory acuity, decision-making, social connectedness, and the like that become more prevalent with age (Fozard et al., 2000). Results from research on aging are used to inform the technical aspects of product design, housing, mobility, information and communications, safety and security systems, training and education, health and home care, and medical technology (Graafmans & Taipale, 1998).

The functions of gerontechnology include *preventing* problems from occurring, *enhancing* personal ability to overcome problems, *compensating* for losses that cannot be overcome by enhancement, assisting with *care* provision where care is needed, and promoting *research* on problems without current solutions (Fozard, Graafmans, Rietsema, Bouma, & van Berlo, 1996). It would seem that gerontechnology could play a central role in reducing the disproportionate vulnerability of older adults in emergencies and disasters and building disaster resilience. The core functions of gerontechnology (prevention, enhancement, compensation, care, and research) have relevance for meeting the needs of older adults within each phase of the emergency management cycle. Gerontechnology methodologies of inquiry have a critical contribution to make to the development of resources to enhance disaster resilience and reduce risk for older adults. This potential is explored in the following sections.

5.5   Prevention/Mitigation

In emergency management, there is increasing recognition of the value of taking actions in the present to preventing foreseeable problems from occurring in the future (WHO, 2007). For example, Hwacha (2005) describes a consultation process with provinces, territories, and stakeholders in Canada aimed at developing a national disaster mitigation strategy, and Henstra and McBean (2005) highlight the human and economic losses that are motivating this paradigm shift. Mitigation and prevention have traditionally involved technology and engineering, generally at the infrastructure level. For example, floodways are built to reduce the risk of flooding, buildings are engineered to withstand earthquakes, and cyclone shelters are built in strategic locations.

In the field of gerontechnology, there has been a conceptually similar prevention and mitigation focus on developing technologies that enable older people to remain in conventional housing in the community (i.e., in single family detached and semi-detached houses, row housing, low- and high-rise apartment blocks) for much longer than might be expected based on their medical conditions. These technologies include walkers, wheelchairs, and electric scooters for those with mobility limitations, portable oxygen systems for persons with pulmonary disease, home dialysis systems, and various types of personal emergency response systems (PERS). This trend has implications for emergency management. Haq, Whitelegg, and Kohler (2008) draw attention to weather-related disasters, expressing concern that the hazards of climate change could negate the value of these technologies without careful planning. In the next section, we highlight some infrastructure-enhancing technologies that have the potential to increase disaster resilience for older adults by mitigating foreseeable risks.

5.5.1   Tracking and Mapping Systems

The two major technologies for tracking and mapping that have been applied to emergency management are global positioning systems (GPS) and geographic information systems (GIS). GPS are satellite-based navigation systems that provide reliable location and time information. GPS technology has become ubiquitous, readily available in vehicles and cell phones. GPS are one of the primary components of computer-facilitated dispatch/response systems. In the event of a disaster, the nearest response units can be selected, routed, and dispatched once the location is known. GIS are database systems that use software to analyze data that can create maps and tables for planning and decision-making (Federal Emergency Management Agency [FEMA], 2011). Use of GIS allows for information sharing of spatial databases (e.g., hydro lines, streets, population distributions) on computer-generated maps. In the United States, GIS are used in emergency management at the federal level by agencies such as FEMA and at the state and local level by police, fire, and other emergency services. The use of Web-based GIS mapping has grown at the organizational level and is also beginning to expand to the personal level (FEMA, 2011). This expansion means people may one day be able to link into an emergency management system via their personal computer to learn how, for example, to evacuate from their town using the best route based on the trajectory of an imminent tornado. Following the 9/11 terrorist attack in the United States, there was a spike in the proposal and development of advanced GIS for emergency management. One example was a proposed GIS-based intelligent emergency response system (GIERS) that would use real-time three-dimensional (3D) GIS (Kwan & Lee, 2005). A 3D system would allow identification of details such as occupancy of a single room in a building that would increase evacuation speed.

GPS and GIS can be used together for a variety of purposes. For example, one California healthcare organization has used the two together to dispatch helicopters and ambulances, to identify the fastest routes for response vehicles, and to interface with police and fire departments and emergency medical services (Hildreth, 2007). Significant efforts are underway to apply these technologies to humanitarian emergencies (Johnson, 2000; Kaiser, Spiegel, Henderson, & Gerber, 2003). It is critical that the distinct needs of frail older adults in community settings and in long-term care facilities are represented in the evolution of these systems as emergency management resources. Applying this technology to emergency management in large-scale disasters, GIS and GPS technology could be used together to improve evacuation of older adults to temporary shelters. Where disaster zones are known to encompass neighborhoods with high concentrations of older adults who are likely to need assistance evacuating from their homes, use of these systems can assist emergency responders to effectively deploy scarce resources. If long-term care facilities are included in the system, both sending and receiving facilities would benefit from increased ability to place older adults appropriately, considering both bed availability and care needs. The combined systems could be used to track persons with dementia and others whose medical conditions impair their ability to self-report their location before, during, and after an evacuation. Systems that include the capacity for advanced vehicle locating (tracking the location of transportation vehicles in real time) would permit response to any untoward situation that might arise during the transportation of vulnerable older adults from one setting to another. If the integrated systems are advanced enough, medical records could also be incorporated within them. This would be of particular value to older adults, who are likely to experience disruptions in medical care for preexisting conditions during disaster situations. Maintaining treatment for common conditions such as arthritis, cardiovascular disease, diabetes, and cancer is often dependent upon timely access to healthcare records that document medical treatment routines. Information sharing between agencies can be difficult due in large part to a strong reliance on paper-based systems or on a computer system where data are not accessible outside the facility. Past disasters in the United States have demonstrated that the lack of an electronic medical record system delays the provision of health services for frail older persons (Hyer et al., 2006), while the availability of such a system can expedite resumption of services (United States Department of Veterans Affairs, 2005).

Technological challenges for the use of GIS and GPS in emergency management can include dropped signals, imprecise mapping, and difficulty linking systems and matching coordinates. Cost-effective and widely accessible monitoring technologies that will allow superior communication and integrated response are needed (Anderson & Gow, 2003). Overarching requirements for the success of this technology as a resource for disaster prevention and mitigation include refining interconnectivity between different systems to guarantee completeness and continuity of information flow between organizations, facilities, transfer locations, and care documentation systems. Laws may need to be revised to permit electronic access and sharing of information between tracking and mapping technologies. However, the potential for these systems to reduce vulnerability for older adults in emergencies and disasters is significant (Kiefer, Mancini, Morrow, Gladwin, & Stewart, 2008; Smith et al., 2009).

5.5.2 "Intelligent Building" Systems

A promising development in gerontechnology is the use of *intelligent building* systems in residential settings such as long-term care facilities, assisted living facilities, senior housing, and retirement communities. Although the definition of an intelligent building has been debated over the years, in general, they use a combination of communication, mechanical, electrical, and safety systems that are integrated to create a space that can improve workplace productivity, promote building energy efficiency, allow for remote operations, and ensure safety (Continental Automated Buildings Association [CABA], 2002; Wong, Li, & Wang, 2005). In theory, emergency management organizations can be linked to the safety system of an intelligent building so that timely information is transmitted in emergency situations to appropriate emergency responders (police, fire department, paramedics). The *intelligent Building Response* (iBR) project at the National Institute of Standards and Technology Building and Fire Research Laboratory in the United States is working on the creation of a standard for real-time transfer of information such as location of building occupants and hazards (e.g., fire) from the intelligent building system to emergency response organizations (National Institute of Standards and Technology, 2005). Currently, most of the information collected by intelligent building systems remains within the building which does not allow emergency responders to fully assess the situation until they arrive on the scene.

Smart houses, comprised of dwellings with automated systems that control temperature, lights, sound systems, surveillance cameras, and so forth, are another example of intelligent buildings that could be linked to emergency management organizations (Tiresias, 2009). While currently the cost of owning and managing a smart home is beyond the reach of many seniors, the potential benefits of this technology for building disaster resilience as well as for other purposes may become accessible to larger segments of the aging population in the future.

5.6   Emergency Preparedness

Mitigation and prevention activities can reduce or eliminate the threat posed by hazardous events; however, not all emergencies can be prevented. Still, earthquakes, tsunamis, ice storms, hurricanes, and other such events are predictable in their likelihood, if not their timing, and large-scale industrial accidents, pandemics, terrorist activities, and other hazards of modern life are likely to continue to occur. Emergency preparedness means being ready, willing, and able to respond to these events. The critical questions for individuals are the following: What are the risks to my safety where I live? What information do I need to have to minimize my risks? Where can I obtain this information? In addition to self-reliance, however, there is a collective responsibility for policies, strategies, and programs aimed at emergency preparedness at the level of organizations, communities, and governments. This multidimensional model for preparedness reflects the population health approach to understanding disaster vulnerability and resilience (Berry & Hutton, 2009). In this section, we consider applications of technology in aid of emergency preparedness for both individuals and communities.

5.6.1   Individual Preparedness: Medical and Assistive Devices

Lack of a steady supply of electricity to ensure medications can be maintained at required temperatures for viability and to power medical equipment such as respiratory and dialysis machines can quickly aggravate chronic diseases common to older age (IASC, 2008). The 2003 blackout in the eastern seaboard of Canada and the United States dramatically illustrated this challenge. A study of New York's healthcare system revealed a surge of hospital emergency room visits secondary to respiratory device failure (mechanical ventilators, positive pressure breathing assist devices, nebulizers, and oxygen compressors) (Prezant et al., 2005). It is essential that personal preparedness plans for older adults include information on how to meet the need for temporary backup electricity if this is required to support lifesaving equipment or medications (Gibson & Hayunga, 2006). The same is true for electric wheelchairs and scooters and other electronic assistive devices. How temporary backup strategies function, how affordable they are for older adults, and how easily and safely they may be used are all questions for the field of gerontechnology.

Everyday assistive devices such as eyeglasses and hearing aids are another essential component in emergency preparedness planning for older adults. Hearing loss, due to presbycusis, is very common in older adults. Close to half of men aged 65 and over and nearly a third of older women in the United States reported hearing difficulties in a 2002 study (Federal Interagency Forum on Aging Related Statistics, 2004). With aging, visual acuity is reduced due to presbyopia, and many seniors need glasses to read. Vision may also be limited by cataracts or macular degeneration, placing the senior at risk for adverse events such as falls. Older adults with severe hearing loss and/or without their hearing aids or batteries may not respond to alarms in emergency situations and, if evacuated, are at risk for communication difficulties. Similarly, evacuees will experience difficulties if eyeglasses are lost or forgotten. Alternate communication strategies and access to temporary assistive devices should be part of the personal emergency preparedness plan put in place by an older adult with sensory impairments. Cost and availability of innovative solutions are challenges for gerontechnologists to consider.

5.6.2 Community Preparedness: Communication and Notification Systems

Increasingly, there are examples of social media playing a role in emergency preparedness (Merchant, Elmer, & Lurie, 2011). For example, aid officials in the Philippines credited social media communications with persuading people to take precautions in advance of an October 2010 typhoon (IRIN, 2010). The increasing numbers of people, including older adults, who use the Internet attests to its potential as a communication tool for notification of emergencies and disasters (Czaja & Lee, 2003). However, it is important not to lose sight of the digital divide. A recent national survey indicated that only 54 % of Americans living with a disability use the Internet, compared with 81 % of adults without a disability. Disability was associated with being older, less educated, and living in a lower-income household (Fox, 2011). At the same time, recent research indicates that social networking sites such as Facebook and LinkedIn are attractive to older adults because they are more likely to be living with a chronic disease and reach out for support online (Madden, 2010). The evolution of this trend as a resilience-building component within emergency preparedness warrants close attention.

It is important to avoid overreliance on one technology for communication and notification, however. Public emergency notifications should be made available through a variety of modes and mediums. Older people need to receive emergency notifications that are appropriate to their needs and in accessible formats (WHO, 2008). In the United States, television, radio, community technology centers, sirens/loudspeakers and, when available, telephones with systems such as a 311 information system or reverse 911 are all types of technology that have been used. These systems are potentially the best source of technology to reach vulnerable populations (Kiefer et al., 2008). Reverse 911 automated systems can call large numbers of homes in a designated area quickly as an aid to emergency notification (Gibson & Hayunga, 2006). Contacting staff and family members during a disaster is a challenge for healthcare facilities, including long-term care homes, due to the limited number of phone lines or cell phones. Families are often concerned about the well-being of their older member and resort to phone calls for verbal updates. There are opportunities for social networking technologies, reverse 911 systems, and related technologies to fulfill the need for communication when the usual lines of communication are not operating or a request has been issued to avoid using phones.

There is also potential for PERS to play a greater role in emergency preparedness. This will be especially true, should these systems succeed in expanding their market share in the future. Currently about 10 % of older individuals in the United Kingdom, and around 1 % in North America, use such devices. Ever since the advent of the first generation of PERS, technological developments have improved the capability of users—older adults, response centers, and first responders—to reach someone in need in the shortest possible time. This was and remains the guiding principle of PERS for seniors when indicated by social and health-related factors such as living alone or having had a recent hospitalization for hip or lower limb surgery (Roush & Teasdale, 1997), or for detecting changes in routine activities of daily living that might be prodromal signs of an impending medical problem (Glascock & Kutzik, 2006).

The next two generations of PERS expanded on the growing need to know where vulnerable older adults live and what are their unique health and social circumstances and to be able to communicate with users who initiate a request for assistance. Currently, fourth generation (4G) PERS devices and services in North America are developing the capability for what is referred to as *reverse alerts*. This refers to response centers or emergency agencies being able to both inform users of an impending disaster—for example, a tornado that has just been seen nearby, rising water, an approaching wild fire, or some biological outbreak causing illness in the community—and confirm that the user has received the alert and is taking protective measures.

This emerging bidirectional communications capability was reported at the 7th World Conference of the International Society for Gerontechnology. One of the authors of this chapter (SH) was commissioned by the PHAC Division of Aging and Seniors to conduct an assessment of the use of PERS as an emergency preparedness and management tool. The analysis of 28 North American PERS companies revealed that (1) PERS communications systems are not generally designed for mass broadcast nor are on-person alert devices usually designed for incoming notices; (2) most PERS systems do not have structural and operational requirements in place to respond to disaster management so that specific groups of older adults could be contacted; and (3) geographic coverage is fragmented: that is, a region may be covered by multiple PERS providers, resulting in even greater difficulty for a local authority to distribute messages (Roush & Gutman, 2010).

While these findings point to the early development of bidirectional 4G PERS capabilities, an inquiry to the Center for Aging Services Technologies (CAST) revealed that only two PERS companies were known to have the capacity for reverse alerts (Alwan, M. Results of CAST report on bidirectional capabilities of U.S. PERS companies, personal communication, May 14, 2010). At the time of the inquiry, one bidirectional system in a long-term care facility near Denver, Colorado, had actually been used to alert residents that a tornado watch was in effect and the residents indicated that they had received the message and were beginning to take appropriate measures. The other system converts text such as a public emergency message displayed on TV into a voice message and transmits it to the in-home device of the user. This company had also just developed an *on the go* feature using GPS for compatible mobile phones to locate and warn people who are away from home and may be in harm's way.

5.7 Emergency Response

Evidence from the response phase of disasters suggests that the needs and resources of the aging population are not well integrated within the guidelines and technical response mechanisms available to the emergency and humanitarian community (IASC, 2008). The UN IASC Cluster System provides an organizing mechanism for system-wide preparedness and capacity for technical response to humanitarian crises. Within this system, age is a crosscutting issue that has been identified as needing more attention (Day, Pirie, & Roys, 2007). Vulnerability risks that need to be addressed include lack of attention to the chronic medical conditions common to older adults (as discussed above) and security issues. These include the need for protection from financial abuse and theft of personal belongings and property during and after evacuation; psychological, physical, and sexual abuse; and other forms of mistreatment and exploitation that may be underestimated as concerns for older adults. Responding effectively to the needs of older adults within the disaster-affected community is often hampered by lack of baseline statistical information on the age distribution and health status of the population prior to an emergency. Data that are collected in response to a disaster are often disaggregated by gender only and do not include sufficient information on age (IASC, 2008). More effective data collection processes are needed to ensure that registration, needs assessments, and morbidity and mortality figures are collected and disaggregated by both age and sex (Day et al., 2007).

5.7.1 Needs Assessment Strategies

There is a need for strategies that use participatory rapid assessment (PRA) techniques (e.g., ranking, resource mapping) to facilitate efforts by older people to self-assess their own needs and capacities when an emergency occurs (HelpAge International, 2000). For efficiency, computer software and Internet-based services should be applied to enhance this data collection and utilization (Kaiser et al., 2003). Research on how to engage older adults in such large-scale technology-based data collection, aggregation, and utilization processes is needed.

Technology-based solutions are also needed to support rapid response by local organizations and relief agencies. A Community Disaster Information System (CDIS) has been developed to enhance relationships between nongovernmental organization (NGO) responders and local agencies and suppliers (Troy, Carson, Vanderbeek, & Hutton, 2008). A pilot test, which was also used to identify ways to improve the CDIS, found it to be beneficial to the NGOs, to the individuals affected by the disaster, and to organizations that provided services and resources. It also encouraged collaboration between the local American Red Cross chapter and the community. Although this system was not created specifically for the older population, it could be adapted to ensure that the users both recognize the potential needs of older adults and have improved access to the resources that are necessary to provide an adequate response to these needs.

5.7.2 Medical Support Strategies

Advanced medical support technologies described under the headings of Telemedicine, Telehealth, and Telecare are systems that integrate various communication and information technologies to perform a specific task: enhancing medical access and support. In mass casualty events, decisions have to be made as to who might live if immediate treatment is received, who might not, and who can wait. Disasters are initially local events, managed by the on-site resources that happen to be present when a situation arises. Where infrastructure allows, resources such as telemedicine and telehealth can be invaluable as a triage aid (Balch, 2008). Telemedicine can provide a visual link between a disaster site and experienced medical personnel at a distant facility, serving as a conduit for transfer of clinical data that can enable treatment to start on site (Chelmsford, 2008). In recent disasters, these technologies have been used in conjunction with mobile communication strategies to supplement local emergency response systems (Turnock, Mastouri, & Jivraj, 2008). Improved outcomes that can result from effective use of telemedicine strategies include a reduction in patient transports (particularly important in disasters when transportation systems are compromised), improved care through better matching of patients to dispatch locations, and synchronization of data between site, dispatch, transport, and receiving center. A survey of 1,801 rural emergency medical services organizations in the United States revealed many had limited resources for managing mass casualty events and a high perceived need for training (Furbee et al., 2006). Telemedicine technologies can both facilitate emergency response training and augment response capacity in the event of an actual emergency.

Where the potential of telemedicine as an emergency response strategy is largely untapped is in the field of chronic disease management. Chronic disease exacerbations comprise a sizable disease burden during disasters (Miller & Arquilla, 2008). A post-Hurricane Katrina study documented challenges in providing services for mental health disorders, diabetes, hypertension, respiratory illness, end-stage renal disease, cardiovascular disease, and cancer (Arrieta, Foreman, Crook, & Icenogle, 2009). Maintaining continuity of medication was the most frequently mentioned problem, with a host of contributing factors including lack of access to pharmacy records and financial barriers. A substantial demand for drugs used to treat chronic medical conditions was identified among Hurricane Katrina evacuees to San Antonio, accounting for 68 % of medications prescribed (Jhung et al., 2007).

As an example, the Ontario Telemedicine Network (Government of Ontario, n.d.b) provides videoconferencing for health professional/patient consultation, thereby improving access to diagnosis and treatment. Its Telehomecare program uses monitoring equipment to link patients with chronic diseases to health professionals to allow better control of chronic diseases such as cardiac failure and diabetes, while an emergency service program uses telemedicine in life-threatening conditions including stroke, critical care, trauma, burns, and mental health crises. Ontario Telehealth (Government of Ontario, n.d.a) provides a phone consultation with a nurse and, if necessary a pharmacist, for health advice and support. It is constantly available (everyday, all hours) and is particularly busy during pandemics. Telehomecare can include community alarm systems, sensors to detect motion or falls, and fire and gas alarms that trigger a warning to a response center (Department of Health, 2005). As these information and communication technologies for medical support expand and become routinely available across healthcare and community settings, their components can be enhanced to improve communication and reach isolated seniors, enabling them to avoid decompensating in a crisis. These technologies have the potential to become significant assets for building disaster resilience among older adults with chronic conditions.

5.7.3 Security Strategies

Survival in the initial stages of a disaster is critically dependent on shelter. Obtaining shelter is one of the six most common problems identified by older people in disasters around the world (HelpAge International, 2000), in part due to misconceptions that extended family, community services, and relief agencies can be counted on to look after older adults (IFRC, 2007). Beyond survival, shelter is necessary for security, personal safety, and protection from the climate and can enhance resistance to ill health and disease. During and after evacuation, there is a need to prevent and respond to the risk of gender-based violence and sexual exploitation. Some of the frail and cognitively impaired older adults housed in the Houston Astrodome following Hurricane Katrina were reported to be victims of robbery and exploitation (Dyer et al., 2006). Temporary shelters should be designed with security in mind (The Sphere Handbook, 2004). The extent to which PERS and other technology could play a role in enhancing shelter security for older adults is largely unexplored. The use of surveillance systems raises concerns about privacy, but older adults' acceptance of in-home monitoring and electronic tracking devices generally supports the trade-off between privacy and safety (Demiris et al., 2004; Landau, Werner, Auslander, Shoval, & Heinik, 2010; Wild, Boise, Lundell, & Fousek, 2008).

A second major security concern is that older displaced persons are often not included in tracing and reunification activities (IASC, 2008). Return, repatriation, and reintegration programs should address the challenge and needs of *unaccompanied old* as energetically as those of unaccompanied children (Day et al., 2007). Finding people potentially affected by a disaster can be a daunting task, especially when searching for older adults with impaired cognition, who may not have the capability to utilize communication tools. Many families could not locate elderly members for days or weeks following Hurricane Katrina due to the lack of a tracking system (Dyer et al., 2006).

Following the 9/11 terrorist attack in New York, many lists of missing people and survivors emerged on the Internet. Although the American Red Cross had a Restoring Family Links program, many people either did not know about this resource or chose to post their search elsewhere. After Hurricane Katrina, the People Finder Project sought to merge all data into one database (Csikszentmihályi, 2010). Data on missing persons were collected from the various web sites by volunteers and a new tool, called the People Finder Interchange Format (PFIF), was used to translate data into a single format that was placed into a central database. The compiled data were then sent to the American Red Cross.

Advancing on this technology, the Google Person Finder was developed following the 2010 Haiti earthquake (Csikszentmihályi, 2010). The Person Finder *scraped* web sites, blogs, and bulletin boards for information, photos, and data, translated it using the PFIF, and sent it to a central database. As part of the People Finder, a widget, or technical code, could be placed on web sites that allowed people to post information or search for missing people. With these advancements in Internet technology, emergency response organizations and service agencies could work together to create a simple tool to collect data on older adults who are missing in a disaster.

5.8   Disaster Recovery

In comparison to other components of emergency management, especially preparedness and response, disaster recovery has received proportionately little attention (Phillips, 2009). The evidence from disasters across the world suggests that recovery processes and supports are insensitive to the specific needs and issues of older people. For older people, disruption to the healthcare infrastructure is a particularly critical issue, as chronic conditions can quickly become acute health crises without sustained treatment and management. Other recovery phase tasks that pose significant challenges for older adults include restoring housing, resettlement, and reestablishing social and economic roles and activities (WHO, 2008).

5.8.1   Reconstruction Strategies

Restoring housing and resettlement is complicated for all people who have been displaced due to a disaster. Longitudinal research in Canada on older adults who lost their homes due to flooding in the province of Quebec revealed persistent negative effects on emotional and social functioning, including mourning for loss of home (Maltais & LaChance, 2007). An analysis of post-disaster reconstruction in

Kobe, Japan, following the 1995 Great Hanshin Earthquake chronicles the practical, social, and emotional challenges encountered by low-income older adults in the years following the disaster as they transitioned through various housing options (Otani, 2010). The recovery and rebuilding of New Orleans following the devastation of 2005 is taking years, and it is likely that many of the more vulnerable segments of the population will be excluded from resettlement in their original communities (Cutter et al., 2006). These evacuees may wish to return to their homes but lack either the financial resources or emotional ability to do so. Affluent homeowners and businesses, including coastal resort properties, have insurance and begin to rebuild quickly. Lower-income neighborhoods with affordable housing, access to public transportation, jobs for less-skilled individuals, and social service networks take longer to reinstate, and the repatriation of displaced residents to these areas may be less of a political and social priority.

Continuity of care from home to temporary accommodation and back to home may be a complicated proposition where smart home technologies and other communication and monitoring systems have provided the environmental support that has enabled a frail older adult to maintain residency in the community rather than in a care facility prior to the disaster. This scenario gives rise to a number of challenges. First, the circumstances encountered in shelters and temporary housing can be particularly disruptive to an older person in fragile health and with compromised mobility (WHO, 2008) with the result that the level of support that was adequate to maintain functioning pre-disaster may no longer suffice post-disaster. Second, the resources to reinstate the environmental supports may be limited, delayed, or totally lacking post-disaster. Third, and perhaps most importantly, this is likely to be a problem for a relatively small proportion of the older adult population and as a result may be low on the priority list of agencies responsible for reconstruction and resettlement.

What is more applicable to a greater number of older adults, particularly low-income renters, is that the disproportionate impact of natural disasters may be attributed in part to the likelihood that they lived in substandard housing in unsafe locations (WHO, 2008). Reflecting the population health approach to understanding vulnerability and resilience, the intersection among health status, economics, and disaster recovery is dramatically illustrated by the lot of older adults and people with functional disabilities after Hurricane Katrina. The New York Times reported that 3 years after the devastation caused by Hurricane Katrina, the majority of people who continued to reside in temporary trailers were older adults and those with mental disorders and physical disabilities (Dewan, 2009). Similarly, economically disadvantaged older adults tended to reside in temporary shelter housing the longest after the earthquake that devastated the Kobe and Osaka areas of Japan in 1995 (Otani, 2010). Reports from the 2011 tsunami and earthquake on Japan's northeastern coast show a similar pattern of delayed recovery for older adults, reflecting a combination of health-related, economic, and social factors (IFRC, 2011).

The reconstruction of housing after disasters has a strong technocratic bias, with an emphasis on safety from a construction and materials standpoint. Unfortunately, projects often fail to address issues at the intersection of housing and livability, including the needs and preferences of the residents for whom the houses are to be built (IFRC, 2001; Twigg, 2002). There is a role for gerontechnology in the field of disaster reconstruction, both with respect to seizing the opportunity to incorporate age-friendly technology within rebuilds and applying gerontechnology methods of inquiry to maximize the fit between proposed new environments and old residents.

For residents of long-term care facilities, short-term recovery issues include restoration of vital services and systems. These may include temporary food, water, and shelter for residents, ensuring they have medical care and prescribed medications, and/or restoring electrical services through emergency generators. Sustained power outages can impact a resident's life-support equipment and have an impact on mobility (e.g., when electric wheelchairs are unusable and elevators are disrupted). Short-term recovery occurs when the immediate threats are halted and both basic services and vital needs are restored. Long-term recovery represents an opportunity to make use of new technologies to enhance the disaster resilience of a facility. The long-term care *Hurricane Summit* sponsored by the Florida Health Care Association in February 2006 (Hyer et al., 2006) was convened to identify issues warranting further coordination between long-term care providers and state and federal emergency operations centers (EOCs). Recommendations included identification and development of computer system enhancements needed to bridge EOC structures with long-term care facilities, development of redundant communication systems, and development of centralized tracking systems.

5.9   Summary and Conclusion

Older adults are at increased risk in emergencies and disasters not because of age per se but because they are more likely to have risk factors for vulnerability, including health problems, increased dependence on healthcare and social services, lower socioeconomic status, and restricted social networks. We have explored the potential for technological solutions to combat this disproportionate vulnerability and increase disaster resiliency for older adults in all phases of emergency management. Prevention-focused technologies that can be implemented to increase resilience to disasters include integrating tracking and mapping systems (GIS and GPS) and intelligent building systems with emergency response services. These resources have the potential to be developed so that they support efforts by first responders to locate and reach older adults who need assistance. Technological advances are needed to improve the function and availability of medical and assistive devices at the personal level, and of communication and notification systems at the community level, under conditions of adversity. Known problems that impact older adults in the recovery stage of a disaster include inadequate needs assessment, insufficient medical support for those with chronic illnesses, and inattention to safety issues including personal security. Technology applications including computer-based assessment tools, remote medical care systems (telemedicine, telehealth, telecare), and computer-assisted monitoring and reunification systems have the potential to contribute to increased disaster resilience for older adults by addressing these problems. In the recovery stage of a disaster, home reconstruction is a dominant issue. There is a need to expand gerontechnology applications, including home adaptations that enable older adults to regain or improve the level of functioning they had in their homes (community or facility based) prior to the disruption that accompanies a disaster.

There is a risk that greater reliance on technology in emergency management can lead to a paradigm where the problems that are most readily addressed by technological innovations receive the lions' share of attention, at the expense of the more difficult vulnerability and response issues (Kiefer et al., 2008). The status quo—disproportionate vulnerability for older adults in emergencies and disasters—is unacceptable. The distinct needs of older adults in community settings, as well as the unique circumstances of older adults who live in long-term care facilities, must be represented in the evolution of technology as an important emergency management resource. Several factors may impact the acceptability, uptake, and demand for technology applications in building disaster resilience for older adults. There is a shortage of outcome studies demonstrating the value of technology for promoting disaster resilience, especially regarding cost-effectiveness and efficiency. Research is needed to quantify the value that disaster technology can hold for older people, both in community settings and in care facilities. For example, although the technology to connect intelligent buildings with emergency management has been under development for some time, the cost to create such a system can be prohibitive and the technological requirements may be more advanced than what a long-term care facility can manage in an already busy day. If not implemented correctly, an intelligent building may also restrict the amount of flexibility a user may have to adjust the system to their changing needs.

There is also a need for educational initiatives that raise awareness of technology's potential benefits as a component of the safety envelope for older adults. For example, among long-term care administrators who have had previous problems using technology, there may be resistance to trying new high-tech devices within their sites (Yu, Li, & Gagnon, 2009). This applies to older adults as well. Simplifying technology for a broad spectrum of users, with varied levels of computer literacy, competence, and comfort, is an ongoing challenge. Much more research on domotics, the field of studying interactions between users and electronic devices, must be conducted (Harrington & Harrington, 2000). And since the older adults of tomorrow will be much more tech-savvy than today's older population, human factors research must be conducted on what various cohorts of end users need and will use.

Practical, relevant research is needed, which takes into account user knowledge and beliefs, the context in which older adults and caregivers interact with given technologies, and the characteristics of the technology itself (Rogers & Fisk, 2010). Designers of technology supports need to work closely with both older adults and other stakeholders to learn how their products and services can be engineered and marketed to promote their use in emergency management. Learning from their feedback would not only improve the design of the technologies but also enhance the processes of technical support. Identification of *best practices* selected from among ongoing efforts, as well as continued planning and implementation of outcome-oriented field pilot tests and larger-scale demonstration projects, is recommended. If we are to expand the technology safety envelope for older adults to include disaster resilience, it will be critical that research on the application of technology for emergency management purposes goes beyond focus groups and self-report surveys to include disaster simulations and exercises, as has been recommended for other assistive technologies (Gutman, 2003).

Age-friendly adaptations to current emergency management practices that would enhance outcomes for older people would in many instances also benefit other population subgroups (IASC, 2006). This is both an argument for greater age sensitivity and inclusiveness in emergency management as a routine practice and an endorsement of the value of technology as a means to this end. Gerontechnology is fundamentally about technological development in context. The potential contribution of technological innovation to disaster resilience will only be realized if it is informed by an understanding of the nature and consequences of human aging and an appreciation of the role and protection responsibilities of communities towards their older members.

References

Abellan Van Kan, G., Rolland, Y., Bergman, H., Morley, J. E., Kritchevsky, S. B., & Vellas, B. (2008). The IANA task force on frailty assessment of older people in clinical practice. *The Journal of Nutrition, Health & Aging, 12*(1), 29–37.

Alzheimer Society of Canada. (2010). *Rising tide: The impact of dementia on Canadian society.* Accessed March 5, 2013 from http://www.alzheimer.ca/en/Get-involved/Raise-your-voice/Rising-Tide

Anderson, P. S., & Gow, G. A. (2003). A general framework for mitigation-oriented planning assessments of mobile telecommunication lifelines (MTL). *Natural Hazards, 28*, 305–318.

Arrieta, M. I., Foreman, R. D., Crook, E. D., & Icenogle, M. L. (2009). Providing continuity of care for chronic diseases in the aftermath of Katrina: From field experience to policy recommendations. *Disaster Medicine and Public Health Preparedness, 3*, 174–182.

Balch, D. (2008). Developing a national inventory of telehealth resources for rapid and effective emergency medical care: a white paper developed by the American Telemedicine Association Emergency Preparedness and Response Special Interest Group. *Telemedicine Journal and E-Health, 14*(6), 606–610.

Barney, C. E., & Roush, R. E. (2009). Emergency preparedness and response considerations for the geriatric population. *Texas Public Health Association Journal, 61*(4), 39–41.

Berry, P., & Hutton, D. (2009). Resilient Canadians, resilient communities. *Health Research Bulletin, 15*, 29–32. Accessed 22 September 2011 from http://www.hc-sc.gc.ca/sr-sr/alt_formats/hpb-dgps/pdf/pubs/hpr-rps/bull/2009-emergency-urgence/2009-emergency-urgence-eng.pdf

Brown, L. M., Hyer, K., & Polivka-West, L. (2007). A comparative study of laws, rules, codes and other influences on nursing homes' disaster preparedness in the Gulf Coast states. *Behavioral Science and Law, 25*(5), 655–675.

Charness, N., & Jastrzembski, T. S. (2009). Gerontechnology. In P. Saariluoma & H. Isomaki (Eds.), *Future interaction design II*. London: Springer.

Chelmsford, N. (2008). *AMD telemedicine explores the role of telemedicine during disasters.* Accessed September 22, 2011 from http://www.amdtelemedicine.com/media/press-080417.html

Continental Automated Buildings Association (CABA). (2002). *Technology roadmap for intelligent buildings.* Accessed September 22, 2011 from http://www.caba.org/trm

Csikszentmihályi, C. (2010, January). *Looking for Haiti's lost, online: How information technology can streamline Web searches.* Columbia Journalism Review. Accessed September 22, 2011 from http://www.cjr.org/the_observatory/looking_for_haitis_lost_online.php?page=all

Cutter, S. L., Emrich, C. T., Mitchell, J. T., Boruff, Gall, B. J., Schmidtlein, M., et al. (2006). The long road home: Race class and recovery from hurricane Katrina. *Environment, 48*(2), 8–20.

Czaja, S. J., & Lee, C. C. (2003). The impact of the internet on older adults. In N. Charness & K. W. Schaie (Eds.), *Impact of technology on successful aging* (Societal impact on aging series). New York, NY: Springer.

Czaja, S. J., Sharit, J., Charness, N., Fisk, A. D., & Rogers, W. (2001). The Centre for Research and Education on Aging and Technology Enhancement (CREATE): A program to enhance technology for older adults. *Gerontechnology, 1,* 50–59.

Day, W., Pirie. A., & Roys, C. (2007). *Strong and fragile: Learning from older people in emergencies.* London: HelpAge International. Accessed March 5, 2013 from http://www.who.int/hac/network/interagency/iasc_advocacy_paper_older_people_en.pdf

Deeg, D. J. H., Huizink, A. C., Comijs, H. C., & Smid, T. (2005). Disaster and associated changes in physical and mental health in older residents. *European Journal of Public Health, 15*(2), 170–174.

Demiris, G., Rantz, M. J., Aud, M. A., Marek, K., Tyrer, H., Skubic, M., et al. (2004). Older adults' attitudes towards and perceptions of 'smart home' technologies: a pilot study. *Medical Informatics and the Internet in Medicine, 29*(2), 87–94.

Department of Health. (2005). *Building telecare in England.* London: Department of Health. Accessed September 22, 2011 from http://www.dh.gov.uk/prod_consum_dh/groups/dh_digitalassets/@dh/@en/documents/digitalasset/dh_4115644.pdf

Dewan, S. (2009). Leaving the trailers: Ready or not, Katrina victims lose temporary housing. *The New York Times.* Accessed September 22, 2011 from http://www.nytimes.com/2009/05/08/us/08trailer.html

Donaldson, G. C., Robinson, D., & Allaway, S. L. (1997). An analysis of arterial disease mortality and BUPA health screening data in men, in relation to outdoor temperature. *Clinical Science, 92*(3), 261–268.

Dosa, D. M., Grossman, N., Wetle, T., & Mor, V. (2007). To evacuate or not to evacuate: Lessons learned from Louisiana nursing home administrators following Hurricanes Katrina and Rita. *Journal of the American Medical Directors Association, 8*(3), 142–149.

Dyer, C., Festa, N., Cloyd, B., Regev, M., Schwartzberg, J. G., James, J., et al. (2006). *Recommendations for best practices in the management of elderly disaster victims.* Houston, TX: Baylor College of Medicine. Accessed September 22, 2011 from http://www.bcm.edu/pdf/bestpractices.pdf

Federal Emergency Management Agency (FEMA). (2011). *Mapping and analysis center.* Accessed March 5, 2013 from http://gis.fema.gov/GISActivities.html

Federal Interagency Forum on Aging Related Statistics. (2004). *Older Americans 2004: Key indicators of well-being.* Washington, DC: US Government Printing office. Accessed September 22, 2011 from http://www.agingstats.gov/Agingstatsdotnet/Main_Site/Data/Data_2004.aspx

Fox, S. (2011). *Americans living with a disability and their technology profile. Pew Internet and American life project.* Accessed September 22, 2011 from http://pewinternet.org/Reports/2011/Disability.aspx

Fozard, J. L., Graafmans, J. A. M., Rietsema, J., Bouma, H., & van Berlo, A. (1996). Aging and ergonomics, the challenges of individual differences and environmental change. In K. A. Brookhuis, C. Weikert, J. Moraal, & D. De Waard (Eds.), *Aging and human factors proceedings of the European chapter of the human factors and ergonomics society annual meeting 1993.* Haren, The Netherlands: Traffic Research Centre, University of Groningen.

Fozard, J. L., Rietsema, J., Bouma, H., & Graafmans, J. A. M. (2000). Gerontechnology: Creating enabling environments for the challenges and opportunities of aging. *Educational Gerontology, 26,* 331–344.

Frohlich, N., De Coster, C., & Dik, N. (2002). *Estimating personal care home bed requirements.* Winnipeg: Manitoba Centre for Health Policy, University of Manitoba. Accessed September 22, 2011 from http://mchp-appserv.cpe.umanitoba.ca/reference/pch2020.pdf

Furbee, P. N., Coben, J. H., Smyth, S. K., Manley, W. G., Summers, D. E., Sanddal, N. D., et al. (2006). Realities of rural emergency medical services disaster preparedness. *Prehospital and Disaster Medicine, 21*(2), 64–70.

Gibson, M. J., & Hayunga, M. (2006). *We can do better: Lessons learned for protecting older persons in disaster.* Washington, DC: AARP. Accessed on September 22, 2011 from http://assets.aarp.org/rgcenter/il/better.pdf

Glascock, A. P., & Kutzik, D. M. (2006). The impact of behavioral monitoring technology on the provision of health care in the home. *Journal of Universal Computer Science, 12*(1), 59–79.

Goodwin, J. (2007). A deadly harvest: The effects of cold weather on older people in the UK. *British Journal of Community Nursing, 12*(1), 23–26.

Government of Ontario. (n.d.a). *Ontario telehealth.* Accessed September 22, 2011 from http://www.health.gov.on.ca/en/public/programs/telehealth/

Government of Ontario. (n.d.b). *Ontario Telemedicine Network.* Accessed September 22, 2011 from http://www.otn.ca

Graafmans, J. A. M., & Taipale, V. (1998). Gerontechnology: A sustainable investment in the future. In J. A. M. Graafmans, V. Taipale, & N. Charness (Eds.), *Gerontechnology: A sustainable investment in the future studies in health technology and infomatics* (Vol. 48). Amsterdam: IOS.

Gutman, G. M. (2003). Gerontechnology and the home environment. In N. Charness & K. W. Schaie (Eds.), *Impact of technology on successful aging* (Societal impact on aging series). New York, NY: Springer.

Haq, G., Whitelegg, J., Kohler, M. (2008). *Growing old in a changing climate: Meeting the challenges of an aging population and climate change.* Stockholm: Stockholm Environment Institute. Accessed September 22, 2011 from http://sei-international.org/mediamanager/documents/Publications/Future/climate_change_growing_old.pdf

Harrington, T. L., & Harrington, M. K. (2000). *Gerontechnology: Why and how.* Maastricht, The Netherlands: Shaker.

HelpAge International. (2000). *Older people in disasters and humanitarian crises: Guidelines for best practice.* London: HelpAge International. Accessed September 22, 2011 from http://www.helpage.org/resources/publications/?adv=0&ssearch=guidelines&filter=f.yeard&type=&region=&topic=&language=&page=4

Henstra, D., & McBean, G. (2005). Canadian disaster management policy: Moving toward a paradigm shift? *Canadian Public Policy, 31*(3), 303–318.

Hildreth, S. (2007). GPS and GIS: On the corporate radar. *Computerworld, 41*(14), 23–26. Accessed September 26, 2011 from http://www.computerworld.com/s/article/284307/On_the_Corporate_Radar

Hutton, D. (2008). *Older persons and emergencies: Considerations for policy and action.* Geneva: World Health Organization. Accessed September 22, 2011 from http://www.who.int/ageing/emergencies/en/index.html

Hwacha, V. (2005). Canada's experience in developing a national disaster mitigation strategy: A deliberative dialogue approach. *Mitigation and Adaptation Strategies for Global Change, 10*, 507–523.

Hyer, K., Brown, L. M., Berman, A., & Polivka-West, L. (2006). Establishing and refining hurricane response systems for long-term care facilities. *Health Affairs, 25*(5), 407–411.

Inter-Agency Standing Committee (IASC). (2006). *Guidance note on using the cluster approach to strengthen humanitarian response.* Accessed March 5, 2013 from http://www.unhcr.org/refworld/docid/460a8ccc2.html

Inter-Agency Standing Committee (IASC) Working Group. (2008). *Humanitarian action and older persons: An essential brief for humanitarian actors.* Accessed September 22, 2011 from http://www.who.int/hac/network/interagency/news/older_persons/en/index.html

International Federation of Red Cross and Red Crescent Societies (IFRC). (2001). *World disasters report 2001: Focus on recovery.* Geneva: IFRC.
International Federation of Red Cross and Red Crescent Societies (IFRC). (2004). *World disasters report 2004: Focus on community resilience.* Geneva: IFRC.
International Federation of Red Cross and Red Crescent Societies (IFRC). (2007). *World disasters report 2007: Focus on discrimination.* Geneva: IFRC. Accessed September 22, 2011 from http://www.ifrc.org/Global/Publications/disasters/WDR/WDR2007-English.pdf
International Federation of Red Cross and Red Crescent Societies (IFRC). (2011). *Japan Diary 5: Taking special care of the sick and elderly.* Accessed September 22, 2011 from http://www.ifrc.org/en/news-and-media/news-stories/asia-pacific/japan/japan-diary-5-taking-special-care-of-the-sick-and-elderly/
IRIN. (2010). *In brief: Social media network helps prevent disaster.* Nairobi: IRIN. Accessed September 22, 2011 from http://www.irinnews.org/PrintReport.aspx?ReportID=90821
Jhung, M. A., Shebab, N., Rohr-Allegrini, C., Pollock, D. A., Sanchez, R., Guerra, F., et al. (2007). Chronic disease and disasters medication demands of Hurricane Katrina evacuees. *American Journal of Preventive Medicine, 33*(3), 207–210.
Johnson, R. (2000). *GIS technology for disasters and emergency management an ESRI white paper.* New York: Environmental Systems Research Institute. Accessed September 22, 2011 from http://www.esri.com/library/whitepapers/pdfs/disastermgmt.pdf
Kaiser, R., Spiegel, P. B., Henderson, A. K., & Gerber, M. L. (2003). The application of geographical information systems and global positioning systems in humanitarian emergencies: Lessons learned, programme implications and future research. *Disasters, 27*(2), 127–140.
Kiefer, J, Mancini, J., Morrow, B., Gladwin, H., & Stewart, T. (2008). *Providing access to resilience-enhancing technologies for disadvantaged communities and vulnerable populations.* Oak Ridge: The Institute for Advanced Biometrics and Social Systems Studies. Accessed September 22, 2011 from http://www.orau.org/university-partnerships/files/The-PARET-Report.pdf
Kosatsky, T. (2005). The 2003 heat wave. *Eurosurveillance, 10*(07). Accessed September 27, 2011 from http://www.eurosurveillance.org/ViewArticle.aspx?ArticleId=552
Kwan, M.-P., & Lee, J. (2005). Emergency response after 9/11: The potential of real-time 3D GIS for quick emergency response in micro-spatial environments. *Computers, Environment and Urban Systems, 29*(2), 93–113.
Laditka, S. B., Laditka, J. N., Cornman, C. B., Davis, C. B., & Richter, J. V. (2009). Resilience and challenges among staff of gulf coast nursing homes sheltering frail evacuees following Hurricane Katrina, 2005 implications for planning and training. *Prehospital and Disaster Medicine, 24*(1), 54–62.
Landau, R., Werner, S., Auslander, G. K., Shoval, G. K., & Heinik, J. (2010). What do cognitively intact older people think about the use of electronic tracking devices for people with dementia? A preliminary analysis. *International Psychogeriatrics, 22*(8), 1301–1309.
Lindsay, J. R. (2003). The determinants of disaster vulnerability: Achieving sustainable mitigation through population health. *Natural Hazards, 28*(2–3), 291–304.
Madden, M. (2010). *Older adults and social media: Social networking use among those ages 50 and older nearly doubled over the past year.* Washington: Pew Research Centre. Accessed September 22, 2011 from http://pewinternet.org/Reports/2010/Older-Adults-and-Social-Media.aspx
Maltais, D., & LaChance, L. (2007). Les conséquences à moyen et à long terme des inondations de juillet 1996 sur la santé biopsychosociale des personnes âgées. *Vie et vieillissement, 6*(2), 23–29.
McGregor, M. J., Tate, R. B., Ronald, L. A., McGrail, K. M., Cox, M. B., Berta, W., et al. (2010). *Staffing in long-term care in British Columbia, Canada: A longitudinal study of differences by facility ownership, 1996–2006.* Statistics Canada: Catalogue no. 82-003-XPE Health Reports. Accessed September 26, 2011 from http://www.statcan.gc.ca/pub/82-003-x/2010004/article/11390-eng.pdf

Merchant, R. M., Elmer, S., & Lurie, N. (2011). Integrating social media into emergency-preparedness efforts. *The New England Journal of Medicine, 365*(4), 289–291.

Miller, C., & Arquilla, B. (2008). Chronic disease and natural hazards: Impact of disasters on diabetic, renal, and cardiac patients. *Prehospital and Disaster Medicine, 23*(2), 185–194.

National Institute of Standards and Technology. (2005, November 7). 'Smart' buildings to guide future first responders. *ScienceDaily.* Accessed September 22, 2011 from http://www.science-daily.com/releases/2005/11/051107080620.htm

Norris, F. H., Stevens, S. P., Pfefferbaum, B., Wyche, K. F., & Pfefferbaum, R. L. (2008). Community resilience as a metaphor, set of capacities, and strategy for readiness. *American Journal of Community Psychology, 41*(1–2), 127–150.

Otani, J. (2010). *Older people in natural disasters.* Kyoto/Melbourne, VI: Kyoto University Press/Trans Pacific.

Phillips, B. D. (2009). *Disaster recovery.* New York, NY: CRC/Taylor and Francis.

Powell, S. (2009). The health impacts of disasters: Who is most at risk? *Health Policy Research Bulletin, 15*, 23–28. Accessed September 22, 2011 from http://www.hc-sc.gc.ca/sr-sr/alt_formats/hpb-dgps/pdf/pubs/hpr-rps/bull/2009-emergency-urgence/2009-emergency-urgence-eng.pdf

Prezant, D. J., Clair, J., Belyaev, S., Alleyne, D., Banauch, G. I., Davitt, M., et al. (2005). Effects of the august 2003 blackout on the New York city healthcare delivery system: A lesson for emergency preparedness. *Critical Care Medicine, 33*(1 Suppl), S96–S101.

Public Health Agency of Canada. (2003). *What makes Canadians healthy or unhealthy?* Ottawa: Public Health Agency of Canada. Accessed September 22, 2011 from http://www.phac-aspc.gc.ca/ph-sp/determinants/determinants-eng.php#evidence

Public Health Agency of Canada. (2008a). *Building a global framework to address the needs and contributions of older people in emergencies.* Ottawa: Minister of Public Works and Government Services Canada. Accessed September 22, 2011 from http://www.phac-aspc.gc.ca/seniors-aines/publications/pro/emergency-urgence/global-mondial/index-eng.php

Public Health Agency of Canada. (2008b). *Second international workshop on seniors and emergency preparedness: Workshop report.* Ottawa: Public Health Agency of Canada. Accessed September 22, 2011 from http://www.phac-aspc.gc.ca/seniors-aines/publications/pro/emergency-urgence/workshop-colloque/index-eng.php

Public Health Agency of Canada. (2010). *The chief public health officer's report on the state of public health in Canada, 2010.* Ottawa: Public Health Agency of Canada. Accessed September 22, 2010 from http://www.phac-aspc.gc.ca/cphorsphc-respcacsp/2010/fr-rc/pdf/cpho_report_2010_e.pdf

Rogers, W. A., & Fisk, A. D. (2010). Toward a psychological science of advanced technology design for older adults. *Journals of Gerontology B Psychological Science and Social Science, 65B*(6), 645–653.

Roush, R., & Gutman, G. (2010). Using preventative gerotechnology systems to monitor residents behavior for health services during emergencies. *Gerontechnology, 9*(2), 92.

Roush, R. E., & Teasdale, T. A. (1997). Reduced hospitalization rates of two sets of community residing older adults before and after use of a personal response system. *Journal of Applied Gerontology, 16*(3), 355–366.

Saliba, D., Buchanan, J., & Kingston, R. S. (2004). Functions and response of nursing facilities during community disaster. *American Journal of Public Health, 94*(8), 1436–1441.

Shields, T. J., Boyce, K. E., & McConnell, N. (2009). The behaviour and evacuation experiences of WTC 9/11 evacuees with self-designated mobility impairments. *Fire Safety Journal, 44*(6), 881–893.

Smith, S. M., Tremethick, M. J., Johnson, P., & Gorski, J. (2009). Disaster planning and response: Considering the needs of the frail elderly. *International Journal of Emergency Management, 6*(1), 1–13.

Statistics Canada. (2010). *Residents on books in residential care facilities by age group, sex, pricipal characteristic of the predominant group of residents and size of facility, Canada, provinces and territories, Annual (Number), 2006.* Accessed September 22, 2011 from http://www5.statcan.gc.ca/cansim/pick-choisir?lang=eng&id=1075504&pattern=1075504 &searchTypeByValue=1

The Sphere Handbook. (2004). *Humanitarian charter and minimum standards in disaster response.* Accessed March 5, 2013 from http://www.unhcr.org/refworld/publisher,SPHERE,,, 3d64ad7b1,0.html

Tiresias. (2009). *What is a smart home?* Accessed September 22, 2011 from http://www.tiresias. org/research/guidelines/smart_home.htm

Troy, D. A., Carson, A., Vanderbeek, J., & Hutton, A. (2008). Enhancing community-based disaster preparedness with information technology. *Disasters, 32*(1), 149–165.

Turnock, M., Mastouri, N., & Jivraj, A. (2008). *Pre-hospital application of telemedicine in acuteonset disaster situations.* Accessed September 22, 2011 from http://www.un-spider.org/sites/ default/files/Prehospital%20telemedicine%20in%20disasters.pdf

Twigg, J. (2002). *Technology, post-disaster housing reconstruction and livelihood security.* Accessed March 5, 2013 from http://practicalaction.org/post-disaster-reconstruction-2

United States Department of Veterans Affairs. (2005). *VA's computerized patient record system saves critical medical records so patients' health care does not suffer.* Accessed 22 September 2011 from http://www.houston.va.gov/pressreleases/News_20050913a.asp

Wild, K., Boise, L., Lundell, J., & Foucek, A. (2008). Unobtrusive in-home monitoring of cognitive and physical health: Reactions and perceptions of older adults. *Journal of Applied Gerontology, 27*(2), 181–200.

Wong, J. K. W., Li, H., & Wang, S. W. (2005). Intelligent building research: A review. *Automation in Construction, 14*(1), 143–159.

World Health Organization. (2007). *Risk reduction and emergency preparedness: WHO six-year strategy for the health sector and community capacity development.* Geneva: World Health Organization. Accessed September 22, 2011 from http://www.who.int/hac/techguidance/preparedness/emergency_preparedness_eng.pdf

World Health Organization. (2008). *Older persons in emergencies: an active ageing perspective.* Geneva: World Health Organization. Accessed September 22, 2011 from http://www.who.int/ ageing/publications/EmergenciesEnglish13August.pdf

Yu, P., Li, H., & Gagnon, M. P. (2009). Health IT acceptance factors in long-term care facilities: A cross-sectional survey. *International Journal of Medical Informatics, 78*(4), 219–229.

Chapter 6
Measuring the Effectiveness of Assistive Technology on Active Aging: Capturing the Perspectives of Users

Jeffrey Jutai and Kenneth Southall

6.1 Introduction

The United States' Assistive Technology Act (1998) defines assistive or adaptive technology as "…any item, piece of equipment, or product system, whether acquired commercially, modified, or customized, that is used to increase, maintain, or improve functional capabilities of individuals with disabilities" (SEC. 3. DEFINITIONS AND RULE subsection 3). This chapter uses the term mobility assistive technology (MAT) to describe a category that includes both assistive devices (applied to or directly manipulated by a person—e.g., a cane, walker, or wheelchair) and special equipment (attachments to the original structure of the physical environment—e.g., grab bars in the bathroom) that are designed to improve mobility.

It is helpful to distinguish the effects of medical and assistive technologies since the latter are especially important in supporting active aging (Jutai et al., 2009a). Medical technologies are more narrowly defined and are designed for assessment and intervention at the level of physical health and healing or, in the language of the International Classification of Functioning, Disability and Health (ICF), *body function and structures* (World Health Organization [WHO], 2001).

J. Jutai, Ph.D., C.Psych. (✉)
Interdisciplinary School of Health Sciences, University of Ottawa, 25 University Street, Ottawa, ON, Canada K1N EN5
e-mail: jjutai@uottawa.ca

K. Southall, Ph.D.
Institut Raymond-Dewar Centre de recherche interdisciplinaire en réadaptation, 3600 rue Berri, Montréal, PQ, Canada H2L 4G9
e-mail: southall.kenneth@gmail.com

A. Sixsmith and G. Gutman (eds.), *Technologies for Active Aging*, International Perspectives on Aging, DOI 10.1007/978-1-4419-8348-0_6, © Springer Science+Business Media New York 2013

Medical devices are not designed to directly and appreciably improve quality of life (QoL) and well-being. These improvements are much more influenced by MAT which helps individuals to engage in life activities and participate in society. Assistive technology provides a platform to support active aging as defined by the WHO. The challenge, however, is to demonstrate the efficacy of particular devices.

6.2   Challenges for Outcomes Measurement

According to the ICF (WHO, 2001), participation refers to the involvement of an individual in a real-life situation. A participation restriction is experienced when, for example, a person has difficulty attending educational classes or other learning activities or engaging in social activities with family, friends, neighbors, or groups. Very commonly, these difficulties result from diminished mobility that occurs naturally with aging or from illness or injury experienced while aging, and many forms of rehabilitation are available to accommodate them. Participation measures, particularly those that measure subjective participation, provide important information for individuals working in the field of rehabilitation since they assess how the individual is doing in the community, which some would argue, is the ultimate rehabilitation outcome. In fact, participation has been termed the most meaningful outcome of rehabilitation (Cicerone, 2004); however, it is probably also the most challenging to measure since there are many things that contribute to a person's level of participation (Dijkers, 2010).

6.3   Does Mobility Assistive Technology Enhance Participation?

This question is more difficult to answer than one might expect. Historically, it was assumed that beneficial effects of assistive technology on participation must be obvious and easily observable, but since they often are not, there is a dearth of studies showing positive outcomes (Fuhrer, Jutai, Scherer, & DeRuyter, 2003). Most existing health and rehabilitation studies have one of three limitations (a) assistive technology impact is not even considered, (b) where it is, there is a failure to recognize that use of assistive technology lowers the functional impact score, and/or (c) the impact is not attributed to specific assistive devices (Rust & Smith, 2005). A review of 20+ years of published assistive technology outcomes research (Lenker, Scherer, Fuhrer, Jutai, & DeRuyter, 2005) revealed several trends in outcome measurement. Device usage and usability accounted for almost 70 % of the dependent variables appearing in the literature, while outcome domains such as functional level, participation, and quality of life accounted for less than 30 %. Of 212 outcome variables used in 82 reviewed articles, 79 % ($N=168$) were measured using non-standardized, study-specific measurement tools with scant reporting of the psychometric properties of the tools used.

As a result of the field's approach to measurement, the assistive technology outcomes research literature is vulnerable to one of three fundamental flaws (a) genuine treatment effects may be undetected because of measures that have weak reliability; (b) tools lacking validity may result in systematic underestimation, overestimation, or misrepresentation of treatment effects; or (c) the impacts of MAT in outcome domains of interest are simply not measured.

Other limitations of measurement tools include narrowly defined outcome domains, lack of comparability of results across instruments, and unacceptable trade-offs between instrument precision and practical implementation. Additionally, many measurement tools cannot be used across the variety of settings in which rehabilitation services are provided, nor along the continuum of care (Jette & Haley, 2005; Jutai, 1999; Jutai, Ladak, Schuller, Naumann, & Wright, 1996; Lenker et al., 2005; Seale & Turner-Smith, 2003). The available pool of traditional measures, in other words, is inadequate to measure outcomes of use of mobility assistive technologies. New tools and techniques are needed. MAT outcomes, including impact on participation as defined by the ICF (WHO, 2001), must be understood and measured within the context of personal choices and activity-level factors. The inherent complexities of this model can only be measured efficiently using dynamic assessment techniques, such as computer adaptive testing (CAT) built on item response theory (IRT) approaches (Gershon, Heinemann, & Fisher, 2006). Input from diverse stakeholder groups is required to form a conceptual basis of MAT outcomes and to assure relevance across a range of disabling conditions. Our research program, described below, addresses these challenges.

6.4 Assistive Technology Outcomes Profile for Mobility

The Assistive Technology Outcomes Profile for Mobility (ATOP/M) is based on state-of-the-art applications of IRT and CAT and designed for evaluating the impact on activities and participation of increased mobility resulting from the use of MATs. With the ATOP/M, the impact can be attributed to specific assistive devices. With CAT, the respondent's ability level relative to a norm group can be iteratively estimated during the testing process and items can be selected based on their current ability estimate. Thus, respondents will receive few items that are very easy or very hard for them. This tailored item selection can result in reduced standard errors and greater precision with only a handful of properly selected items.

The targeted populations for the ATOP/M are adults who have a mobility disability, as defined by the ICF (WHO, 2001) that limits their ability to move from place to place independent of any form of assistance. The ATOP/M is applicable to all relevant conditions that produce this disability and to all settings in which mobility devices may be used, including transfers to/from mobility devices.

An important feature of our research is that it has been very sensitive to the perspectives of MAT users. The ATOP/M includes information that users believe would assist them in better adopting and using an outcomes measure. It presents items in ways that users find understandable and sensible.

6.5 Materials and Methods

The first step in developing the instrument was to assemble an item pool. A total of 148 items were drawn from the Patient-Reported Outcomes Measurement Information System (PROMIS, 2008), the Community Participation Indicators (CPI, v. 4) from the Rehabilitation Research and Training Center (A. Heinemann, personal communication, January 21, 2008), and over a dozen other instruments (Jutai et al., 2009b). Items were mapped to the ICF to ensure adequate coverage of relevant domains. Gaps in content were identified, and items were added from the other instruments. Nine focus groups were then conducted in the USA and Canada with multiple stakeholder groups to identify additional items that were needed and to refine item wording and response format (Hammel et al., 2012). The participants included 45 mobility device users, 10 caregivers of device users, and 10 individuals involved in delivering mobility device services. MAT users had a wide range of diagnoses, including cerebral palsy, multiple sclerosis, spinal cord injury, acquired brain injury, stroke, and amputation. Mobility devices used included canes (quad, single point, side), walkers (wheeled, side), wheelchairs (manual and powered), scooters, prosthetic legs, and crutches. The focus group discussions were analyzed thematically. Items were subjected to binning and winnowing by an expert panel and then cognitively tested with 20 device users.

The ATOP/M pool was field tested in a web-based survey of MAT users to calibrate items. A total of 1,037 subjects were recruited from a national data registry. Sample characteristics were as follows: the age range was 16–95 years, 41 % were male and 81 % were white. Canes (32 %), power (18 %) and manual (13 %) wheelchairs, and walkers (14 %) were the most commonly used devices. Users had a wide range of conditions responsible for their mobility difficulties, including cerebral palsy, multiple sclerosis, spinal cord injury, acquired brain injury, stroke, and amputation. Examples of ATOP/M items are presented below.

Rasch analyses were conducted to create full-length and short-form instruments, and an item bank for CAT applications was developed. The criterion for determining the reliability of the measures, separation reliability, was a value of 0.80, a value that distinguishes at least three distinct strata of respondents (high, average, and low) and a relatively wide range of task frequency (expected to be at least 4 log-odd units) for the items. The internal validity of the scale was assessed using fit statistics. External validity was assessed by comparing the ATOP/M domain and subdomain measures for groups of respondents who differed in the type of assistive device they used. External validity was assessed by comparing the ATOP domain and subdomain measures for groups of respondents who differed in the type of assistive device they used.

The full-length ATOP/M consists of 68 items distributed across two domains, each having two subdomains: Activities (Physical Performance and Instrumental Activities of Daily Living) and Participation (Social Role Performance and Discretionary Social Participation). The naming of the domains and subdomains is consistent with the taxonomy published by the PROMIS project (Cella et al., 2010;

Hahn et al., 2010). The Activities domain is defined as one's ability to carry out various activities that require physical capability, ranging from self-care (activities of daily living) to more vigorous activities that require increasing degrees of mobility. The ATOP/M items have a *capability* stem and a corresponding *capability* set of response items (e.g., "Are you able to…normally, without any difficulty, with a little difficulty, with some difficulty, with much difficulty, unable to do") and are given in the present tense. The Activities domain has two related subdomains: Physical Performance (mobility and lower extremity function) and ability to carry out Instrumental Activities of Daily Living. The Participation domain is defined as involvement in one's usual social roles in life's situations and activities. It includes two subdomains that describe *social roles* such as work and family responsibilities and more *discretionary social activities* such as leisure activity and relationships with friends.

Respondents are asked to answer each item under two different conditions—with device and without device. The instructions are presented as follows:

With device

For the following questions, you will be asked to think about how you usually do the activity, *with the use of a device*. This device can be the primary one that you identified during the previous part of the interview or any of the secondary mobility devices you identified.

Without device

You will now be asked to answer the same series of questions but this time thinking about your ability to do the activity *without* any *device*.

The following are example items.

Activities

Physical Performance: Are you able to get around your neighborhood or town?
Instrumental Activities of Daily Living: Are you able to do yard work like raking leaves, weeding, or pushing a lawn mower?

Participation

Discretionary Social Participation: Are you able to participate in active recreational activities?
Social Role Performance: Are you able to go to classes or participate in learning activities?

The ATOP/M yields two scores, one reflecting respondents' mobility level while using a device and the other reflecting their capability without it. As anticipated, the domain and subdomain measures were lower when respondents rated task difficulty when not using the device than when using the device. All item banks had acceptable reliability, were essentially unidimensional, and had acceptable model fit for most items. Evidence of validity was found in significant differences for persons using canes or scooters compared with those using manual or power wheelchairs (i.e., the unaided performance of wheelchair users was poorer than that of cane and scooter users).

Some of the items proposed by the focus groups did not survive the psychometric analyses. Although they had meaningful content for users, scores on these items did not have distributions and other statistical properties that were adequate for scale construction. These items concerned the effect of MAT on helping people to manage their disability:

Are you able to... use different options for getting around?
Are you able to... get support from other people who have disabilities to help you manage your disability?
Are you able to... get the services you need to better manage your disability?
Are you able to... get the information you need to better manage your disability?
Are you able to... use the Internet to find information that you need to better manage your disability?

In proposing these items for the ATOP/M, focus group members described how being effective at management meant that they could use their MAT to participate in the face of significant challenges in their physical and social environments. In comparison, individuals who could not manage well reported that they abandoned MAT or gave up choice and control in life participation (Hammel et al., 2012). Some focus group participants described using many different types of MAT, conceptualizing them as sets or integrated solutions. They pointed to a need to evaluate MAT outcomes in a way consistent with this reality rather than focusing on a single piece of technology.

6.6   Discussion

The ATOP/M comprises items that reflect activities and participation in areas that are important to users of mobility assistive devices. Items cover a wide range of functional ability and reflect most categories of the ICF framework. The instrument has excellent person and item reliability and no significant ceiling and floor effects. The ATOP/M has been translated into Canadian-French, and its validity is being further examined in several Canadian studies. A similar approach to instrument research and development is under way in Europe and has shown promising results (Brandt, Kreiner & Iwarsson, 2010; Brandt et al., 2008). It includes items derived from the ICF framework and has used Rasch analysis for scale construction.

The development of the ATOP/M revealed that end-users have some expectations for the impact of mobility assistive technology that are not easily measured. Assistive technology outcome assessment tool developers should consider ways to address these expectations. It is also useful to employ a variety of methodologies. For example, visual qualitative methodologies are increasingly being employed to investigate issues related to health care (e.g., Riley & Manias, 2004), and these has been successfully used with older adults (Lockett, Willis, & Edwards, 2005; Magilvy, Congdon, Nelson, & Craig, 1992). One such method, which we are currently using to better understand user perspectives, is called Photovoice. Photovoice is a technique used by researchers and community members to represent their community (Wang & Burris, 1997) and to assess needs and study health promotion topics (Wang, 1999). In this technique, cameras are provided to elderly MAT users who are then asked to take photographs that portray, illustrate, symbolize, or represent their lives. Researchers meet with the photographers to discuss the meanings of the images that were captured, to understand life as seen by the photographers (Hurworth, 2003). The purpose of asking participants to collect photographs is not to recreate or to illustrate empirical truths or *reality*; rather, photographs are primarily used as a medium of communication between researcher and participant about the impact of MAT. The use of photographs facilitates the involvement of elderly research participants and makes them feel more comfortable discussing issues that may be socially stigmatizing. Although the photographs may not contain new information per se, they may inspire alternate meaning-making concerning participation that may otherwise remain dormant in a face-to-face interview (Prosser & Schwartz, 2004).

The broader context for assistive technology and active aging includes recognizing when the need is not for functional assistance but for policy change. Many of the factors that influence device adoption and use are directly or indirectly associated with stigma. Assistive technology use may potentially draw unwanted attention to the user, resulting in embarrassment and fear of being identified as "disabled" (Kaplan, Grynbaum, Rusk, Anastasia, & Gassler, 1966; Kochkin, 2007; Mann, Hurren, & Tomita, 1993). Cultural considerations such as societal attitudes about disability and the value placed on integrating individuals who use assistive technology into society play a critical role in device adoption and use (Scherer, Jutai, Fuhrer, Demers, & Deruyter, 2007). MAT may be a very helpful prophylactic intervention for some elders who are experiencing a decline in physical capability. Why should they wait until after they have had a serious fall, and related injury, before adopting MAT? Measuring the effectiveness of assistive technology on active aging should include the impact of public perceptions of device use on the behaviors of elders.

6.7   Conclusions

The ATOP/M is a conceptually grounded instrument for measuring the outcomes of mobility device use. The advances represented in this instrument are a clarified conceptual model and precise measurement of outcomes by adaptively administering only questions that retrieve maximum information from the device user, thus minimizing respondent burden. While the ATOP/M successfully measures the performance of activities and participation that are important and meaningful to MAT end-users, it does not address all MAT-related issues that are a concern to end-users and may impact active aging. MAT users appear to prioritize a broader range of functional activity, participation, and social/societal outcomes than is typically captured in questionnaires. This suggests the need for a tool or an item pool that includes this range of outcomes and considers differentially weighting the value and importance of items.

MAT users appear to distinguish functional activity and participation from the broader societal impact and value associated with active aging. New and imaginative approaches to outcomes measurement are needed.

References

Brandt, A., Kreiner, S., & Iwarsson, S. (2010). Mobility-related participation and user satisfaction: construct validity in the context of powered wheelchair use. *Disability and Rehabilitation: Assistive Technology, 5*(5), 305–313.

Brandt, Å., Löfqvist, C., Jónsdottir, I., Sund, T., Salminen, A.-L., Werngren-Elgström, M., et al. (2008). Towards an instrument targeting mobility-related participation: Nordic cross-national reliability. *Journal of Rehabilitation Medicine, 40*, 766–772.

Cella, D., Riley, W., Stone, A., Rothrock, N., Reeve, B., Yount, S., et al. (2010). The patient-reported outcomes measurement information system (PROMIS) developed and tested its first wave of adult self-reported health outcome item banks: 2005–2008. *Journal of Clinical Epidemiology, 63*, 1179–1194.

Cicerone, K. D. (2004). Participation as an outcome of traumatic brain injury rehabilitation. *The Journal of Head Trauma Rehabilitation, 19*(6), 494–501.

Dijkers, M. P. (2010). Issues in the conceptualization and measurement of participation: an overview. *Archives of Physical Medicine and Rehabilitation, 91*(9), S1–S76.

Fuhrer, M. J., Jutai, J. W., Scherer, M. J., & DeRuyter, F. (2003). A framework for the conceptual modeling of assistive technology outcomes. *Disability and Rehabilitation, 25*, 1243–1251.

Gershon, R., Heinemann, A. W., & Fisher, W. P. (2006). Development and application of the orthotics and prosthetics user survey: applications and opportunities for health care quality improvement. *Journal of Prosthetics & Orthotics, 18*, 80–85.

Hahn, E. A., DeVellis, R. F., Bode, R. K., Garcia, S. F., Castel, L. D., Eisen, S. V., et al. (2010). Measuring social health in the patient-reported outcomes measurement information system (PROMIS): item bank development and testing. *Quality of Life Research, 19*, 1035–1044.

Hammel, J., Southall, K., Jutai, J., Finlayson, M., Kashindi, G., & Fok, D. (2012). Evaluating use and outcomes of mobility technology: A multiple stakeholder analysis. *Disability and Rehabilitation: Assistive Technology*, Nov 9 [Epub ahead of print].

Hurworth, R. (2003). Photo-interviewing for research. *Social Research Update, 40*, 1–4.

Jette, A. M., & Haley, S. M. (2005). Contemporary measurement techniques for rehabilitation outcomes assessment. *Journal of Rehabilitation Medicine, 37*, 339–345.

Jutai, J. (1999). Quality of life impact of assistive technology. *Rehabilitation Engineering, 14*, 2–7.

Jutai, J. W., Coulson, S., & Russell-Minda, E. (2009a). In Amichai-Hamburger (Ed.), Technology and psychological well-being. Cambridge: Cambridge University Press, pp. 206–226.

Jutai, J. W., Demers, L., DeRuyter, F., Finlayson, M., Fuhrer, M. J., & Hammel, J. (2009b, June). *Assistive technology outcomes profile for mobility (ATOP/M)–item pool development*. New Orleans, LA: Rehabilitation Engineering and Assistive Technology Society of North America (RESNA).

Jutai, J., Ladak, N., Schuller, R., Naumann, S., & Wright, V. (1996). Outcomes measurement of assistive technologies: An institutional perspective. *Assistive Technology, 8*, 110–120.

Kaplan, L. I., Grynbaum, B. B., Rusk, H. A., Anastasia, T., & Gassler, S. (1966). A reappraisal of braces and other mechanical aids in patients with spinal cord dysfunction: Results of a follow-up study. *Archives of Physical Medicine and Rehabilitation, 47*, 393–405.

Kochkin, S. (2007). MarkeTrak VII: Obstacles to adult non-user adoption of hearing aids. *The Hearing Journal, 60*, 24–51.

Lenker, J. A., Scherer, M. J., Fuhrer, M. J., Jutai, J. W., & DeRuyter, F. (2005). Psychometric and administrative properties of measures used in assistive technology device outcomes research. *Assistive Technology, 17*, 7–22.

Lockett, D., Willis, A., & Edwards, N. (2005). Through seniors' eyes: An exploratory qualitative study to identify environmental barriers to and facilitators of walking. *The Canadian Journal of Nursing Research, 37*, 48–65.

Magilvy, J., Congdon, J., Nelson, J., & Craig, C. (1992). Visions of rural aging: Use of photographic method in gerontological research. *The Gerontologist, 32*, 253–257.

Mann, W. C., Hurren, D., & Tomita, M. (1993). Comparison of assistive device use and needs of homebased older persons with different impairments. *The American Journal of Occupational Therapy, 47*, 980–987.

PROMIS Health Organization and PROMIS Cooperative Group. (2008). PROMIS Item Pool v.1.0. Retrieved from http://www.nihpromis.org

Prosser, J., & Schwartz, D. (2004). Photographs within the sociological research process. In S. Hesse-Biber & P. Leavy (Eds.), *Approaches to qualitative research: A reader on theory and practice* (pp. 334–349). New York, NY: Oxford University Press.

Riley, R. G., & Manias, E. (2004). The uses of photography in clinical nursing practice and research: A literature review. *Journal of Advanced Nursing, 48*, 397–405.

Rust, K., & Smith, R. O. (2005). Assistive technology in the measurement of rehabilitation and health outcomes: A review and analysis of instruments. *American Journal of Physical Medicine & Rehabilitation, 84*(10), 780–793.

Scherer, M., Jutai, J., Fuhrer, M., Demers, L., & Deruyter, F. (2007). A framework for modelling the selection of assistive technology devices (ATDs). *Disability and Rehabilitation: Assistive Technology, 2*, 1–8.

Seale, J. K., & Turner-Smith, A. R. (2003). Measuring the impact of assistive technologies on quality of life: can rehabilitation professionals rise to the challenge? In A. J. Carr, I. J. Higginson, & P. G. Robinson (Eds.), *Quality of life*. London: BMJ Books.

United States' Assistive Technology Act (1998). Assistive Technology Act of 1998. Retrieved April 5, 2012 from http://www.section508.gov/508Awareness/html/at1998.html

Wang, C. C. (1999). Photovoice: A participatory action research strategy applied to women's health. *Journal of Women's Health, 8*, 185–192.

Wang, C., & Burris, M. A. (1997). Photovoice: Concept, methodology, and use for participatory needs assessment. *Health Education & Behavior, 24*, 369–387.

World Health Organization. (2001). *International classification of functioning, disability and health (ICF)*. Geneva: WHO.

Chapter 7
Using Technology to Support People with Dementia

Gail Mountain

7.1 Introduction

The aim in this chapter is to challenge the commonly held assumption that once a person develops dementia, they are no longer able to benefit from technology in a full range of life domains—for example, for health, for entertainment, and for leisure—as in addition to the commonly acknowledged role of assistance with safety and security. For a number of reasons, the wider potential of technology is, as yet, relatively unconsidered and unexplored for people with a dementia diagnosis. It is argued that full recognition of the benefits technology could bring to people living with this progressive, irreversible condition is long overdue.

7.2 Background

The impact that technology is having upon all aspects of our lives including work, leisure, and recreation is already significant and is set to escalate in the very near future with developments such as fast speed broadband, cloud computing, and mobile devices of ever increasing sophistication. However, some sectors of society are missing out on this technological explosion. This can be attributed to a number of factors including lack of access due to geographical location, financial limitations, or unmet need for education and support to use and maintain the technology (Hill, Benyon-Davies, & Williams, 2008).

G. Mountain (✉)
Rehabilitation and Assistive Technologies Group, School of Health and Related Research, University of Sheffield, 30 Regent Court, Sheffield S1 4DA, UK
e-mail: g.a.mountain@sheffield.ac.uk

A. Sixsmith and G. Gutman (eds.), *Technologies for Active Aging*, International Perspectives on Aging, DOI 10.1007/978-1-4419-8348-0_7, © Springer Science+Business Media New York 2013

Older people are one group who are less likely to take advantage of new technology in comparison to other population groups despite the emergence of *silver surfers*. Efforts are now being made to facilitate the digital inclusion of this sector of society, for example, through promotional advertising and educational opportunities. It is also recognized that helping older people to maintain community connectedness by remaining integrated in social life is essential for health and well-being and that this can be achieved through new technologies (Ballantyne, Trenwith, Zubrinich, & Corlis, 2010). However, despite the increasing numbers of older people in society, and the associated escalation in needs for health and social care, the specific requirements of older people tend not to be embraced by those developing and marketing technologies.

Those older people who develop dementia are even more likely to be excluded from technological advances even though they could arguably benefit as much if not more than the rest of the population (Astell, 2009). Little effort is being made to get this group connected, and use of commonly encountered technologies to promote autonomy, independence, and enjoyment at all stages of the dementia trajectory is a neglected dimension. A number of specialist devices have started to emerge in recent years with the aim of ameliorating the impact of the illness and assisting people to cope as well as possible. These include devices to assist with cognition, rehabilitation, and functional abilities and to compensate for deficits (Bharucha et al., 2009; LoPresti, Mihailidis, & Kirsch, 2004), but the full benefits that technology might deliver to people with dementia and their carers are not yet exploited. Given the limited consideration of how technology (both everyday and specialist devices) can be used to assist those living with dementia, it is surprising there has been little research to examine the potential benefits (Bharucha et al., 2009). This is particularly so for those with early-stage dementia living in the community (Topo, 2007). As dementia progresses, people become increasingly dependent on family and formal care providers to assist in meeting their needs. Therefore, a balance must be stuck between the needs of people with dementia whose condition means that they become less able to articulate their views, and those of caregivers who have a significant role to play. It is notable that technology developers tend to ask carers for their views regarding needs for, and the nature of, specialist technologies to compensate for the problems arising from the illness in preference to asking those with the diagnosis (Lauriks et al., 2007; Topo, 2007). Also, most studies have been conducted in residential rather than community settings, and only a small proportion are mainstreamed in practice.

Recognizing that people with the condition and their carers have their own distinct and individual needs and views is critical in designing technology that can assist people with the diagnosis as well as those who provide care for them. Accepting that these needs and views may differ is an important part of this process and will allow us to go beyond safety and security to consider how technology might help individuals to encompass the things that make life worth living, such as accomplishment, engagement, and meaningful interactions with others. It is these issues that this chapter focuses upon.

7.3   Dementia in Society

Much has been written about the increasing numbers of older people globally and what this means for individuals, for societies, and for populations. One of the well-documented consequences of the aging population is the dramatic increase in the numbers of people with dementia.

Dementia is an umbrella term used to describe a range of conditions characterized by progressive cognitive decline and associated changes in communication, memory and perception, all of which undermine abilities for independent living (National Institute for Clinical Excellence and Social Care Institute for Excellence, 2006). The trajectory of decline is highly variable from person to person and is also dependent upon with which specific condition the individual is diagnosed. Symptoms can be unpredictable creating significant challenges for the person when trying to manage their illness on a day-to-day basis which can be further complicated by changing levels of insight (Aalten, Van Valen, Clare, Kenny, & Verhey, 1995). Age is the greatest risk factor for developing dementia, and as the population ages, the number of people with dementia is predicted to rise. Current estimates in the UK put the number of people with some form of dementia at approximately 820,000 (Luengo-Fernandez, Leal, & Gray, 2010) with this number expected to increase to 940,000 million by 2021 (Alzheimer's Society UK, 2010). A similar prevalence exists in Canada, where it is estimated that there are around 500,000 people living with Alzheimer's disease or another form of dementia. This translates into approximately one in 11 people over 65 and one in three of those over 85 being affected. These projections are echoed globally, with the prediction being that numbers of people with dementia within the world population will double every 20 years to 81.1 million by 2040 (Ferri et al., 2005). The costs of care are expected to rise concomitantly from recent estimates of £17 billion per annum (Alzheimer's Society UK, 2010). This includes both the direct costs of providing care and services to individuals and the indirect costs of informal caregiving, which incorporates loss of workforce productivity by those with caring roles. Rising numbers of those affected by various types of dementia and the extent and complexity of needs they present are leading to escalating demands for health and social care with consequent significant resource implications (MacDonald & Cooper, 2007) at local, national, and international levels (Ferri et al., 2005).

Population statistics and projections do not always convey the heterogeneity that exists among those who are diagnosed with dementia. They extend from those in midlife with early-onset dementia to those who are very old. Some people with dementia may have other health conditions, especially those who are already quite old when they are diagnosed. Additionally, individuals who develop dementia come from diverse cultural and social backgrounds, living circumstances, and networks of care. Thus needs differ from person to person and it is difficult to predict longer-term outcome. The resulting complexity of needs is one of the reasons for what can be inadequate service responses. Most people with a dementia diagnosis live at home for most of the course of their illness, typically cared for by family members

(Luengo-Fernandez et al., 2010). As the person's condition worsens, their reliance on other people increases, often resulting in additional demand on health and social care services as their relatives struggle to cope (Schneider, Murray, Banerjee, & Mann, 1999).

7.4   Responding to the Needs Arising from Dementia

Societal and service responses to the needs of people with dementia are improving, but there is a significant legacy of inadequate provision. The neglect of the needs of people with dementia can be attributed to a combination of factors such as the previously described complexity of needs, past history of service provision within countries, the associated poor levels of funding for services, and patchy provision across geographical areas within the same region or country. The combination of low societal perceptions combined with limited service development has led to a self-fulfilling prophecy of minimal expectations, denial of opportunities, and resultant loss of functioning (Graff et al., 2006).

Until very recently, dementia was a hidden topic of concern, rarely acknowledged by policymakers, the media, or society in general. This marginalization is illustrated by the lack of investment in services specializing in dementia diagnosis, treatment, and care (Waldemar et al., 2007). It is demonstrated by inadequate management of people who have a dementia diagnosis in general hospital settings (Harwood, Hope, & Jacoby, 1997). There has also been widespread neglect of training and education for staff working within services where people with dementia are likely to present.

Late diagnosis can occur for two main reasons. Firstly, people who are experiencing symptoms of cognitive impairment can be fearful of the consequences of diagnosis so do not seek it (Bond et al., 2005). Secondly, physicians and, in particular, general practitioners may be reluctant to diagnose (Downs, Clibbens, Rae, Cook, & Woods, 2002). These problems can be attributed, in part, to the lack of treatment options that used to exist; until recently diagnostic confirmation was interpreted as meaning that nothing more could be done for the individual. However, the increasing availability of medication (cholinesterase inhibitors) intended to enhance memory function has proved to be a powerful force for change. As well as providing some improvements in cognitive function for a period of time (Birks, 2006), the availability of this form of medication is changing perceptions of the range and extent of help that can be offered to people. As a consequence people are beginning to present earlier and clinicians can be more inclined to make the diagnosis.

As previously mentioned, caregivers are frequently instrumental in enabling the person to remain in the community rather than entering residential care. It is therefore not surprising that service provision post-diagnosis used to focus almost exclusively upon the needs of family carers so that they are equipped to continue caring.

Fig. 7.1 Downward spiral: self-fulfilling prophecy

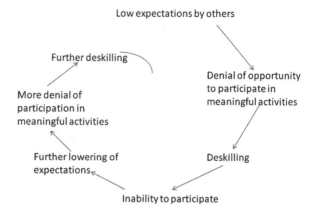

This could have the effect of accelerating loss of skills on the part of the person with dementia as illustrated in the diagram (Fig. 7.1).

The most recent understanding is that people with dementia can live a quality life and can be helped to retain skills and abilities for as long as possible. This involves helping the individual to identify strategies to limit the impact of cognitive deficits upon functional abilities and providing interventions such as cognitive training and cognitive rehabilitation so that skills can be retained for longer (Kurz, Pohl, Ramsenthaler, & Sorg, 2009).

Self-management by people with early- to middle-stage dementia is being increasingly accepted as being a real possibility (Moniz-Cook & Vernooij-Dassen, 2006; Mountain, 2006). Self-management requires individuals to understand their condition and use strategies to cope with their symptoms, so that independence can be retained for as long as possible. It requires giving priority to the views and needs of the person with dementia while taking account of their social and personal circumstances and the views of others, such as family members, who are directly affected by the illness. Taking steps to ensure continued physical and mental well-being is integral to successful self-management, but goals must be realistic and not require a lifestyle or skill level that may be unattainable or unsustainable (Mountain, 2006). The self-management approach emphasizes the need to strike a balance between the needs of people with dementia and those of family members fulfilling a caregiving role (Mountain & Craig, 2012). It is in these domains in particular that appropriate technology can make a significant contribution.

These changes in perceptions of how people with dementia might be assisted to live a quality life are illustrated by the following statement from the UK Social Care Institute for Excellence derived from the results of a practice-based project to pilot the use of computers in day centers for people with dementia:

> *...people with dementia can engage in computer work at many different levels...the common misconception that people with dementia cannot learn new skills or be able to benefit from being involved in interacting with computers ... is untrue.*

Social Care Institute for Excellence: Good Practice Example 01

A key element in changing societal perceptions of the abilities and needs of people with dementia is the increasing number who are speaking up for themselves. Interestingly technology, particularly the Internet, has played a part in this as reflected in the growth of blogs and wikis created by people with dementia as illustrated by the following excerpts:

> *I have taken up photography...it seems to be one of the few things I am still pretty good at doing. This is our baby robin on his/her first day out of the nest...I love the little pin feathers saying 'I'm not quite finished yet...' just like me. I am not finished yet either even though I have recently been diagnosed with early onset dementia.*

> *These poems especially help me to deal with my situation today. It helps me to put into perspective where I am, who I am, what I am. Reading these poems helps me to define the strengths I do have. The promise. It allows me to put behind me those unkind, unthinking people who do not understand dementia...and me.* (Quotes extracted from Health Central. com, documenting dementia, 2008)

Another example is provided through the Internet-based Dementia Advocacy and Support Network (DASNE) created by and for people with dementia which has been in existence for several years.

This changing awareness of the abilities of people with dementia including the increasing advocacy on the part of people with the condition is the backdrop against which technologies are being introduced and mainstreamed. Nevertheless, to date the bulk of technological applications for dementia has focused on addressing safety and security issues, largely in response to the concerns of caregivers, with much less attention paid to meeting other, and we would argue equally if not more important needs such as communication and social participation. A review of how technology is being developed and appropriated for people with dementia is provided in the next sections of this chapter.

7.5   Technologies to Promote Health and Safety

Radical new solutions are sought by policymakers to manage the impact of aging and long-term conditions upon services. Technology is perceived to be an important means of delivering treatment and care within the constraints of health and social care budgets (DoH, 2010). A new model of healthcare delivery has been identified that does not focus upon acute care and is underpinned by information and communication technologies (European Commission Information Society and Media, 2006). Within this model, technology to deliver interventions that promote health and well-being is promoted as a highly satisfactory means of meeting the needs of people with long-term conditions in individual, cost-effective ways (Liddell, Adshead, & Burgess, 2008). As a consequence, both service commissioners and providers are being encouraged to adopt the use of Telecare and Telehealth technologies for a range of long-term conditions.

7.5.1   Telecare

Telecare is remote monitoring to manage risks, thereby assisting the individual to live independently (Brownsell & Bradley, 2003). There has been a significant policy push in the UK to introduce Telecare within social care services (DoH, 2006). The majority of Telecare devices on the market are concerned with managing the functional consequences of illness and disability and, in the case of people with dementia, tend to compensate for failing memory and poor orientation, with a focus upon promotion of safety. Increasingly, people with dementia are being viewed as ideal recipients of Telecare technology; the UK National Dementia Strategy (DoH, 2009) has an objective dedicated to housing and Telecare.

Examples of Telecare include community alarms to raise assistance, sensors to monitor movement around the home, technology to remind the individual about tasks in progress and prevent accidents such as leaving cookers on and taps running, and devices to provide surveillance outside the home. As well as technologies being stand alone, devices can be hardwired into the fabric of the house, creating a *smart* home environment (Martin, Kelly, Kernohan, McCreight, & Nugent, 2008). Many established Telecare devices such as community alarms and sensors in the home were not originally designed for people with dementia but are increasingly being offered as solutions to help a person with cognitive impairment to remain within the community. However, the evidence base and in particular the social and financial benefits have yet to be adequately demonstrated. Mainstreaming Telecare will produce better and cheaper solutions, but infrastructure requirements are a significant consideration (British Society of Gerontology, 2009).

Given the previously described changing views of how people with dementia can be assisted to retain skills and independence for as long as possible, it is disappointing to observe that much of Telecare deployment to date remains concerned with providing carer reassurance (Astell, 2006) rather than meeting the needs of the individual with the diagnosis. This reflects outdated modes of service delivery outlined in the previous section where carer needs were prioritized over those of the person with the condition. It also may be a reflection of the fact that many Telecare devices have not been specifically designed to meet the needs of people with dementia. Some work has been conducted to look at the specific needs of people with dementia and how they might be met through Telecare; for example, ASTRID (a social and technological response to meeting the needs of individuals with dementia and their carers) was a European Union project that explored this. The project identified four key principles for providing Telecare for people with dementia. These are (1) autonomy, people should be able to decide what they want to happen or be done to them; (2) beneficence, we should try to do good to the people we care for; (3) nonmaleficence, we should try to avoid doing people harm; and (4) justice, people should be treated fairly and equally (Marshall, 2000).

In the UK most Telecare equipment provision demands some degree of professional prescription, but this will change with the introduction of a retail model for assistive technologies through the UK Department of Health initiative *Transforming*

Community Equipment Services. This means that people in the UK will be able to purchase certain technology from high street shops, whereas previously it could only be obtained from specialist outlets and should lead to better access, improved design, and more competitive pricing.

The complex ethical issues underpinning the use of technology with people with dementia and in particular Telecare technologies are only just beginning to be addressed. It has been suggested that innovations to enable people to be independent while at the same time tracking their location dehumanize the person with dementia by restricting their liberty and failing to take account of their views (Hughes & Louw, 2002). Drawing upon the experiences of researchers working in Europe and North America, Mahoney et al. (2007) identified the potentially conflicting views of different stakeholders and how they can manifest. For example, technology developers can set aside the humanity of the person they are creating a device for, those involved in reviewing technology research for ethical approval might provide a biased opinion, and there is also a risk of family carers having unrealistic expectations of the technology so that the person with dementia is left at risk. They pose a framework for what they term a humanistic and ethical approach towards this form of research. The Nuffield Council on Bioethics (2009) has also produced guidance for those involved in commissioning and providing services for people with dementia including assistive technology.

7.5.2   Telehealth

Another form of assistive technology, Telehealth is concerned with the remote delivery of health interventions through technology (Curry, Trejo, & Wardle, 2003). Telehealth is often confused with Telemedicine, which is use of technology for medical diagnosis and consultations. In contrast, Telehealth is about active intervention delivery. Examples include use of the telephone or videophone to provide support to those who are trying to change their health behaviors, for example, to provide remote assistance with the maintenance of exercise for those with heart failure (Paradis, Cossette, Frasure-Smith, Heppell, & Guertin, 2010). Another example of Telehealth is specially designed technology for vital signs monitoring by people in their own homes (Varnfield et al., 2011), and an increasing number of devices are becoming available to enable these functions, for example, the Bosch Health Buddy http://www.bosch-telehealth.com/content/language1/html/55_ ENU_XHTML.aspx. As a result, Telehealth devices are being increasingly incorporated within healthcare pathways for people with long-term conditions and particularly for those with diabetes, chronic heart failure, and chronic obstructive pulmonary disease.

Despite the recognition that people with dementia frequently have to live with other comorbidities, there is a lack of documented evidence regarding use of Telehealth technologies with people with dementia for the purposes of managing comorbidities and few practice examples within the UK. The only current references to use of Telehealth in the context of dementia are concerned with meeting needs resulting from the dementing illness, for example, for providing carer support (Glueckauf, Ketterson, Loomis, & Dages, 2004) or to provide specific assistance with medication management (Smith, Lunde, Hathaway, & Vickers, 2007). This neglect of other needs such as physical health is a concern that is articulated by people with dementia (Mountain & Craig, 2012).

7.6 Technology Development to Meet the Specific Needs of People with Dementia

The previous brief sections on Telehealth and Telecare technologies illustrate the neglect of the specific needs of people with dementia within the majority of mainstreamed technologies. However, there is a developing body of work to develop specific technologies that might benefit people with dementia and their carers. A range of innovations to assist people are in progress, many of which have involved people with dementia and their carers during all stages of technology development which means that there is more likelihood of these devices meeting user requirements.

7.6.1 Involving People with Dementia in Device Development

Ensuring that the needs of the intended user are met entails involving the intended recipients from the outset. The direct involvement of people with dementia in the research process presents challenges for researchers and technology developers. To be adopted, devices must be easy to use, reliable, affordable, and robust. The needs of the intended consumer should inform device development from the outset and must be an integral part of the entire process (Cahill, Macijauskiene, Nygard, Faulkner, & Hagen, 2007). This is demanding but the benefits can be significant and can completely change researcher-constructed design concepts; for example, Sixsmith, Gibson, Orpwood, and Torrington (2007) consulted with people with dementia to identify their needs and wishes which also took into account the practical issues that people were encountering in their daily life. The resultant ideas were used to develop prototype technologies for leisure and recreation. Robinson, Brittain, Lindsay, Jackson, and Olivier (2009) used participatory design methodologies with people with dementia to gain their perspectives on how technologies might support safe walking and independence in the community. This involved obtaining views from people with dementia (to provide a scope of the issues), successfully engaging individuals in participatory design workshops and then in prototype development. The conclusions drawn by the researchers involved in both of these projects were that people in the mild to moderate stages of the disease are able to provide meaningful feedback. Specific issues that have to be taken into account include undertaking consultation in surroundings that are familiar to those being interviewed. The use of visual prompts during consultation can also be invaluable in enabling participants to remain engaged in the process and to compensate for memory problems (Mountain & Craig, 2012). In common with other examples of good practice in engaging end users in technology development, allowances have to be made for those who only want to be involved in a limited way as well as for non-attendance due to illness and dropout. As Adlam, Faulkner, Orpwood, Macijauskiene, and Budraitiene (2004) confirm, establishing and maintaining this level of engagement requires specific skills which technology developers need to nurture. Alternatively, technologists need to work within a multidisciplinary research team where skills in relating to people with dementia already exist. Table 7.1 provides examples of technology from projects that have involved people with dementia in the development process. As well as serving to illustrate the range of needs that innovations are seeking to address, it also serves to underscore the potential for involving people with dementia even in the later stages of the illness.

Many of the technologies shown in Table 7.1 address two problems that people diagnosed with dementia face: difficulty with initiating tasks and difficulty staying on track once a task is started. For example, the work of Wherton and Monk (2009) was concerned with investigating how to prompt a person with moderate dementia to use the cooker dials correctly and safely while cooking. It involved working with care home residents, supported by care workers who they knew well. Other examples in Table 7.1 go beyond the achievement of functional tasks. These include promoting communication, interaction, and relationship building. The INDEPENDENT project (Orpwood et al., 2010) was concerned with developing technology to improve the quality of life of people with dementia. The previously described wish list identified by people with dementia led to a number of prototype devices. The most successful was the CD player for independent use by people with moderate to severe dementia. As the authors describe, some of the design challenges were significant, particularly given that seemingly trivial design decisions could result in confusion and inability to use the player.

7.6.2 Using Existing Technologies to Assist with Independent Living

There is much potential in the application of existing, readily available technologies to the needs of people with dementia as described in the practice guide produced by a UK third sector organization, Dementia Voice (2004). The guide, developed in consultation with people with dementia and their carers, describes how everyday technologies can be used to promote safety and provide reminders together with a consideration of the inherent issues such as ethics. Table 7.2 suggests how existing technologies might be mapped onto three examples of self-management domains for people with dementia (Mountain & Craig, 2012).

Table 7.1 Examples of technologies developed to meet the needs of people with dementia

What technology is intended for	Stage of dementia	Project title/description	Reference to work	Stage of technology development
Prompting activity sequencing during ADL	Moderate/late	Coach device to monitor and prompt hand washing	Mihailidis, Boger, Canido, and Hoey (2007)	Prototype
Prompting safe usage of domestic equipment	Moderate	Prompting/cueing when using the cooker	Wherton and Monk (2009)	Prototype
Following recipes	Early/moderate	Sequencing and prompting	Serna, Pigot, and Rialle (2007)	Prototype
Orientation at night/prompting	All	Nocturnal	Wang et al. (2010)	Prototype
Orientation and targeted prompting	Moderate	SMART home technology to provide prompts	Evans, Orpwood, Adlam, Chadd, and Self (2007); Orpwood et al. (2008)	Prototype
MobileDay navigator	Early	Personalisable mobile technology that prompts daily activities, maintains social contacts and promotes safety	Mulvenna and Nugent (2010)	Prototype
Meaningful activity; pleasure	Early to late	Music player	Orpwood et al. (2010)	Ready for commercialization
Meaningful and engaging activity; autonomy, competence and control	Early to late	Living in the moment—games and activities for people with dementia	Astell et al. (2009, 2010)	Ready for commercialization

Table 7.2 Examples of existing technologies mapped against three self-management domains identified by people living with dementia

Domain	Elements	Examples of existing technologies which might assist
Activities of daily living	(1) In-home instrumental activities of daily living (IADL); cooking, cleaning, gardening, using phone; home security (2) Community-based IADL; using transport, shopping, managing finances (3) Home and community safety	Commercially available safety devices—fire alarm/carbon monoxide alert Commercially available reminder systems Lost item locator Video phone Smartphone technology Combined mobile phone/ GPS technology Door alarm
Keeping connected	(1) Maintaining community connectedness (2) Accessing outside opportunities (3) Using transport	Video phone Reminder systems Internet social networking computer Webcam Combined mobile phone/ GPS technology
Keeping physically well	(1) Managing medication (2) Eating and nutrition (3) Sleep	Medication reminders/dispensers Automatic reminders on fridge, freezer, etc. Commercially available daybreak alarm clock Motion-activated lighting

Another technology that is becoming increasingly ubiquitous in society, the Internet, offers increasing opportunities for people with dementia and their carers. Lauriks et al. (2007) identified four domains in which people with dementia can benefit from ICT (including Internet use). These are information provision, support with managing the symptoms of dementia, provision of opportunities for social interaction, and monitoring for health and safety. The latter domain is of course linked to the provision of Telecare considered above. However, it should also be acknowledged that people with dementia can experience difficulties with using everyday technology as described through the taxonomy developed by Nygard and Starkhammar (2007). Furthermore the tendency to focus exclusively upon carer needs is being extended to exploration of the potential that the Internet might offer, for example, use of webcam, email, Internet, and social networking to support carers (Powell et al., 2010).

7.6.3 Using Existing Technologies to Enable Leisure, Recreation and Enjoyment

The majority of the previously described technological applications tend to be used to assist with the undertaking of daily living activities. However, in common with the rest of society, technology also offers opportunities for enjoyment and socialization for people with dementia, for example, through mobile telephones, computers, cameras, video, and tape recorders. For families living away from relatives, these technologies can form an important link, such as this example from a daughter living remotely from her mother with dementia:

> *Although I've lived in the States for the last two years I don't think that I've ever felt quite so close to Mum. No expectations, no pressures and a sense that throughout it all we were connected, doing it together.*

(Killick & Craig, 2012)

Clearly the communication and staying in touch potential of Internet and mobile technologies is more relevant for people with dementia living in their own home, but increasingly these technologies are also becoming available in long-term care settings. The opportunities offered by existing technology, both for new learning and as a way of enabling people with dementia to demonstrate their skills, are illustrated through the experience of James McKillop. James is a person with dementia who took up photography after receiving a late diagnosis for early dementia. In his book he describes how photography provides him with a powerful medium for self-expression, illustrating this with his images.

As McKillop (2003) wrote in his book:

> *I wanted to raise awareness about dementia and show that people with dementia can re-learn forgotten skills as well as learn something new.*

7.7 Discussion

One aim of this chapter was to illustrate how societal views of dementia are inextricably linked to the forms of technology that are recommended for those living with the condition. The way that some specialist devices mirror the outdated view that people are unable to make any meaningful contribution post-diagnosis is particularly revealing and sadly still lingers in many areas of service provision. However, there are also significant new forms of technological development which move beyond devices whose main function is to provide carer reassurance to those that enable a person with dementia to retain independence and self-efficacy and enjoy life. This is an exciting new area for technology developers and is one that offers great scope as brief illustration of recently prototyped devices indicates.

Also emphasized is the need to work with people with dementia so that devices meet their changing needs. Addressing ethical requirements and concerns is another significant area of consideration for both developers and those engaged in the subsequent introduction of devices. This chapter also draws attention to evidence from research and from practice that demonstrates that people with dementia can and should be helped to use everyday technologies and also be encouraged to benefit from the Internet. Use of existing technology reduces the temptation to place the focus on the technology rather than on the person. In this way technology is seen as a tool and the means for engagement rather than an end in itself, although the initial appeal of a technological tool or gadget may sometimes offer the starting point for an interaction. Furthermore, engagement with everyday as well as specialist technologies can assist individuals to make the most of the skills they retain.

A number of recommendations can be identified out of the bodies of literature and information used for this chapter if people with dementia are to be enabled to benefit from technologies in common with the rest of society. Firstly all those involved in technology for people with dementia, whether in a technology development, commissioning, or providing role, must take time to understand the complexity of dementia. This includes acknowledging the individual needs of people with the condition as well as the needs of their carers. Resulting solutions have to match the requirements of the individual and ideally be personalized. Secondly the direct involvement of people with dementia in the entire research trajectory from needs identification to testing in practice is essential if devices that people need and want are to be made available. This necessitates the application of specific skills on the part of researchers so that people with dementia are engaged in optimum ways. The benefits can be significant and can completely change researcher-constructed design concepts. The third recommendation stems out of the need to appreciate the complexity of dementia and the essential requirement for user engagement. Best practice in engagement of people with dementia within technology development demands multidisciplinary research spanning both technical and clinical expertise so that the necessary skills are available. Industry also has a crucial role to play within this many-faceted research.

Finally more thought needs to be given to the information needs of people with dementia and their carers. A number of UK initiatives are now in place including fact sheets on assistive technology through the UK Alzheimer's Society and a dedicated website to provide details of what is available and where it can be obtained from. However, individual information needs differ and also the time when people are most receptive to receiving it. Thus, there are several considerations here including how to make people with dementia and their carers aware of what is available and how to present information in the most accessible formats.

7.8   Conclusions

This chapter has only been able to touch upon the exciting but increasingly complex debates concerned with technology and people with dementia. Over time there will be an ever increasing range of everyday technologies that are capable of meeting needs and providing leisure opportunities, even in the later stages of the condition. This may negate the requirement for specialist devices and also promote the ability of people to engage with devices rather than being passive recipients of technology-led services.

References

Aalten, P., Van Valen, E., Clare, L., Kenny, G., & Verhey, F. (1995). Awareness in dementia: A review of clinical correlates. *Aging & Mental Health, 9*(5), 414–422.

Adlam, T., Faulkner, R., Orpwood, J. K., Macijauskiene, J., & Budraitiene, A. (2004). The installation and support of internationally distributed equipment for people with dementia. *IEEE Transactions on Information Technology in Biomedicine, 8*(3), 253–257.

Alzheimer's Society UK. (2010). *Demography: Alzheimer's Society position statement.* London: Alzheimer's Society. http://www.alzheimers.org.uk

Astell, A. J. (2006). Personhood and technology in dementia. *Quality in Ageing, 7*(1), 15–25.

Astell, A. J. (2009). REAFF: A framework for developing technology to address the needs of people with dementia. In M. Mulvenna, A. Astell, H. Zheng, & T. Wright (Eds), *Proceedings of the First International Workshop RSW-2009*, Cambridge, UK, CEUR Workshop Proceedings. http://sunsite.informatik.rwth-aachen.de/Publications/CEUR-WS/Vol-499/paper02-Astell.pdf

Astell, A. J., Alm, N., Gowans, G., Ellis, M., Dye, R., & Vaughan, P. (2009). Involving older people with dementia and their carers in designing computer-based support systems: Some methodological considerations. *Universal Access in the Information Society, 8*(1), 49–59.

Astell, A. J., Ellis, M. P., Bernardi, L., Alm, N., Dye, R., Gowans, G., et al. (2010). Using a touch screen computer to support relationships between people with dementia and caregivers. *Interacting with Computers, 22*, 267–275.

Ballantyne, A., Trenwith, L., Zubrinich, S., & Corlis, M. (2010). 'I feel less lonely': What older people say about participating in a social networking site. *Quality in Ageing and Older Adults, 11*(3), 25–35.

Bharucha, A. J., Anand, V., Forlizzi, J., Dew, M. A., Reynolds, C. F., Stevens, S., et al. (2009). Intelligent assistive technology applications to dementia care: Current capabilities, limitations and future challenges. *American Journal of Geriatric Society, 17*(2), 88–102.

Birks, J. (2006). Cholinesterase inhibitors for Alzheimer's disease. *Cochrane Database of Systematic Reviews*, (1), CD005593.

Bond, J., Stave, C., Sganga, A., Vincenzino, O., O'Connell, B., & Stanley, R. L. (2005). Inequalities in dementia care across Europe: Key findings of the facing dementia survey. *International Journal of Clinical Practice, 59*, 8–14.

British Society of Gerontology (2009, September). *National Dementia Strategy: Report of Symposium at the British Society and Gerontology Annual Conference.* http://www.google.co.uk/search?hl=en&q=british+society+of+gerontology+and+national+dementia+strategy&meta=&rlz=1I7ADRA_enGB406

Brownsell, S., & Bradley, D. (2003). *Assistive technology and telecare: Forging solutions for independent living.* Bristol: Policy.

Cahill, S., Macijauskiene, J., Nygard, A. M., Faulkner, J. P., & Hagen, I. (2007). Technology in dementia care. *Technology and Disability, 19*, 56–60.

Curry, R., Trejo, T. M., & Wardle, D. (2003). *Telecare: Using information and communication technology to support independent loving by older, disabled and vulnerable people.* London: Report to the UK Department of Health.

Dasne: Dementia advocacy and services network. http://www.dasne.international.org

Dementia Voice. (2004). *At home with AT (assistive technology).* http://www.housinglin.org.uk/_library/Resources/Housing/Housing_advice/At_Home_with_AT_-_Dementia_Voice_December_2004.pdf

DoH. (2006). *Preventative technology grant 2006–07 to 2007–08.* London: Department of Health.

DoH. (2009). *Living well with dementia: A national dementia strategy.* London: Department of Health.

DoH. (2010). *Generic long term conditions model.* http://www.dh.gov.uk/en/health care/Longtermconditions/DH_120915

Downs, M., Clibbens, R., Rae, C., Cook, A., & Woods, R. (2002). What do general practitioners tell people with dementia and their families about the condition? *Dementia, 1*(1), 47–58.

European Commission Information Society and Media. (2006). *ICT for Health and i2010; Transforming the European healthcare landscape towards a strategy for ICT for Health.* Luxembourg: Author.

Evans, N., Orpwood, R., Adlam, T., Chadd, J., & Self, D. (2007). Evaluation of an enabling smart flat for people with dementia. *Journal of Dementia Care, 15*, 33–36.

Ferri, C. P., Prince, M., Brayne, C., Brodaty, H., Fratiglioni, L., Ganguili, M., et al. (2005). Global prevalence of dementia: A Delphi consensus study. *Lancet, 366*(9503), 2112–2117.

Glueckauf, R. L., Ketterson, T. U., Loomis, J. S., & Dages, P. (2004). Online support and education for dementia caregivers: Overview, utilization, and initial program evaluation. *Telemedicine Journal and e-Health, 10*(2), 223–232.

Graff, M. J., Vernooij-Dassen, M. J. M., Thijssen, M., Dekker, J., Hoefnagels, W. H. L., & Olde-Rikkert, M. G. M. (2006). Community based occupational therapy for patients with dementia and their care givers: A randomised controlled trial. *British Medical Journal, 333*(7580), 1196.

Harwood, D. M. J., Hope, T., & Jacoby, R. (1997). Cognitive impairment and medical in-patients. II: Do physicians miss cognitive impairment? *Age and Aging, 26*, 37–39.

Healthcentral.com. (2008). Documenting dementia. Retrieved April, 2011 from http://www.healthcentral.com/alzheimers/c/6509/19796/dementia-2008

Hill, Benyon-Davies, P., & Williams, M. D. (2008). Older people and internet engagement. *Information Technology and People, 21*(3), 244–266.

Hughes, J. C., & Louw, S. J. (2002). Electronic tagging of people with dementia who wander. Ethical considerations are possibly more important than practical benefits. *British Medical Journal, 325*, 847.

Killick, J., & Craig, C. (2012). *Creativity and communication in persons with dementia.* London: Jessica Kingsley.

Kurz, A., Pohl, C., Ramsenthaler, M., & Sorg, C. (2009). Cognitive rehabilitation in patients with mild cognitive impairment. *International Journal of Geriatric Psychiatry, 24*(2), 163–168.

Lauriks, S., Reinersmann, A., Van der Roest, H. G., Meiland, F. J. M., Davies, R. J., Moelaert, F., et al. (2007). Review of ICT-based services for identified needs in people with dementia. *Ageing Research Reviews, 6*, 223–246.

Liddell, A., Adshead, S., & Burgess, E. (2008). *Technology in the NHS: Transforming the patient's experience of care.* London: The King's Fund.

LoPresti, E. F., Mihailidis, A., & Kirsch, N. (2004). Assistive technology for cognitive rehabilitation: State of the art. *Neuropsychological Rehabilitation, 14*(1–2), 5–39.

Luengo-Fernandez, R., Leal, J., & Gray, A. (2010). *Dementia 2010: The economic burden of dementia and associated research funding in the United Kingdom.* http://www.dementia2010.org/reports/Dementia2010Full.pdf

MacDonald, A., & Cooper, B. (2007). Long-term care and dementia services: An impending crisis. *Age and Ageing, 36*, 16–22.

Mahoney, D. F., Purtilo, R. B., Webbe, F. M., Alwan, M., Bharucha, A. J., Adlam, T. D., et al. (2007). In-home monitoring of persons with dementia: Ethical guidelines for technology research and development. *Alzheimer's & Dementia, 2*, 217–226.

Marshall, M. (Ed.). (2000). *A guide to using technology within dementia care.* London: Hawker.

Martin S., Kelly G., Kernohan W. G., McCreight B., & Nugent C. (2008). Smart home technologies for health and social care support. *Cochrane Database of Systematic Reviews*, (4), CD006412.

McKillop, J. (2003). *Opening shutters—opening minds.* Stirling: Dementia Services Development Centre.

Mihailidis, A., Boger, J., Canido, M., & Hoey, J. (2007). The use of an intelligent prompting system for people with dementia. *ACM Interactions, 14*(4), 34–37.

Moniz-Cook, E., & Vernooij-Dassen, M. (2006). Editorial: Timely psychosocial intervention in dementia: A primary care perspective. *Dementia: The International Journal of Social Care, 5*, 307–315.

Mountain, G. A. (2006). Self management and dementia: An exploration of concepts and evidence. *Dementia: The International Journal of Social Research and Practice, 5*(3), 429–447.

Mountain, G., & Craig, C. (2012). What should be in a self management programme for people with early stage dementia? *Aging & Mental Health.* doi:10.1080/13607863.2011.651430.

Mulvenna, M., & Nugent, C. D. (2010). *Supporting people with dementia using pervasive health technologies*. Berlin: Springer.

National Institute for Clinical Excellence and Social Care Institute for Excellence. (2006). *Dementia: Supporting people with dementia and their carers in health and social care* (NICE clinical guideline 42). London: NICE and SCIE.

Nuffield Council of Bioethics. (2009). *Dementia: Ethical issues*. http://www.nuffieldbioethics.org/dementia

Nygard, L., & Starkhammar, S. (2007). The use of everyday technologies by people with dementia living alone: mapping out the difficulties. *Aging & Mental Health, 11*(2), 144–155.

Orpwood, R., Adiamt T., Evans, N., Chado. J., & Self, D. (2008). Evaluation of an assisted living smart home for someone with dementia. *Journal of Assistive Technologies, 2*(2), 13–21.

Orpwood, R., Bjørneby, S., Hagen, I., Mäki, O., Faulkner, R., & Topo, P. (2004). User involvement in dementia product development. *Dementia, 3*(3), 263–279.

Orpwood, R., Chadd, J., Howcroft, D., Sixsmith, A., Torrington, J., Gibson, G., et al. (2010). Designing technology to improve quality of life for people with dementia: User–led approaches. *Universal Access in the Information Society, 9*, 249–259.

Paradis, V., Cossette, S., Frasure-Smith, N., Heppell, S., & Guertin, M. C. (2010). The efficacy of a motivational nursing intervention on self-care in heart failure patients. *The Journal of Cardiovascular Nursing, 25*(2), 130–141.

Powell, J., Gunn, L., Lowe, P., Sheehan, B., Griffiths, F., & Clarke, A. (2010). New networked technologies and carers of people with dementia: An interview study. *Ageing and Society, 30*, 1073–1088.

Robinson, L., Brittain, K., Lindsay, S., Jackson, D., & Olivier, P. (2009). Keeping in touch everyday (KITE) project: Developing assistive technologies with people with dementia and their carers to promote independence. *International Psychogeriatrics, 21*(3), 494–502.

Schneider, J., Murray, J., Banerjee, S., & Mann, A. (1999). Eurocare: A cross national study of co-resident spouse carers for people with Alzheimer's disease: I—factors associated with carer burden. *International Journal of Geriatric Psychiatry, 14*, 651–661.

Serna, A., Pigot, H., & Rialle, V. (2007). Modeling the progression of Alzheimer's disease for cognitive assistance in smart homes. *User Modeling and User-Adapted Interaction, 17*, 415–438.

Sixsmith, A. J., Gibson, G., Orpwood, R. D., & Torrington, J. M. (2007). Developing a technology based wish list to enhance the quality of life of people with dementia. *Journal of Gerontechnology, 6*(1).

Smith, G. E., Lunde, A. M., Hathaway, J. C., & Vickers, K. S. (2007). Telehealth home monitoring of solitary persons with dementia. *America Journal of Alzheimer's Disease and Other Dementias, 22*(1), 20–26.

Social Care Institute for Excellence: Good Practice Example 01—Improving access to ICT in adult social care settings. http://www.scie.org.uk/workforce/getconnected/examples/example01/index.asp

Topo, P. (2007). Technology studies to meet the needs of people with dementia and their caregivers. *Journal of Applied Gerontology, 18*(1), 5–37.

Varnfield, M., Karunanithi, M. K., Särelä, A., Garcia, E., Fairfull, A., Oldenburg, B. F., et al. (2011). Uptake of a technology-assisted home-care cardiac rehabilitation program. *Medical Journal of Australia, 194*(4), S15–S18.

Waldemar, G., Phung, K. T. T., Burns, A., Georges, J., Hansen, F. R., Iliffe, S., et al. (2007). Access to diagnostic evaluation and treatment for dementia in Europe. *International Journal of Geriatric Psychiatry, 22*, 47–54.

Wang, H., Zheng, H., Augusto, J.C., Martin, S., Mulvenna, M., Carswell, W., et al. (2010). Monitoring and analysis of sleep pattern for people with early dementia. In 1st Workshop on Knowledge Engineering, Discovery and Dissemination in Health (KEDDH'10). *Proceedings of 2010 I.E. International Conference on Bioinformatics and Biomedicine Workshops* (pp. 405–410), Hong Kong, China.

Wherton J., & Monk A. (2009, April). *Choosing the right knob*. Boston, MA: ACM.

Chapter 8
Ambient Assisted Living Systems in Real-Life Situations: Experiences from the SOPRANO Project

Ilse Bierhoff, Sonja Müller, Sandra Schoenrade-Sproll, Sarah Delaney, Paula Byrne, Vesna Dolničar, Babis Magoutas, Yiannis Verginadis, Elena Avatangelou, and Claire Huijnen

8.1 Introduction

The challenge of an aging population requires innovative approaches to meet the needs of increasing numbers of older people within society (Sixsmith & Sixsmith, 2008). In particular, there is a need to move from a health and social agenda that emphasizes dependency to one that promotes active aging and creates supportive

I. Bierhoff (✉) • C. Huijnen
Stichting Smart Homes, Duizelseweg 4A – 5521 AC Eersel, Postbus 8825 – 5605
LV Eindhoven, The Netherlands
e-mail: i.bierhoff@smart-homes.nl; c.huijnen@smart-homes.nl

S. Müller
empirica Gesellschaft für Kommunikations- und Technologieforschung mbH,
Oxfordstr, 2-53111 Bonn, Germany
e-mail: Sonja.Mueller@empirica.com

S. Schoenrade-Sproll
Fraunhofer IAO, Nobelstraße 12, 70569, Stuttgart, Germany
e-mail: sandrasproll@gmx.de

S. Delaney
Work Research Centre, 3 Sundrive Road, Kimmage, Dublin 12, Ireland
e-mail: s.delaney@wrc-research.ie

P. Byrne
Division Primary Care, University Liverpool, Waterhouse Building, Block B, 1st Floor,
1-5 Brownlow Street, L69 3GL, Liverpool, Merseyside, UK

V. Dolničar
Faculty of Social Sciences, University of Ljubljana, Ljubljana, Slovenia
e-mail: Vesna.Dolnicar@fdv.uni-lj.si

B. Magoutas • Y. Verginadis
Institute of Communication and Computer Systems, National Technical University of Athens,
Iroon Polytechniou 9, 15780 Zografou Campus, Athens, Greece
e-mail: elbabmag@mail.ntua.gr; jverg@mail.ntua.gr

A. Sixsmith and G. Gutman (eds.), *Technologies for Active Aging*,
International Perspectives on Aging, DOI 10.1007/978-1-4419-8348-0_8,
© Springer Science+Business Media New York 2013

environments to enable healthy aging in the settings where older people live (Sixsmith et al., 2010). Emerging information and communication technologies (ICTs), such as pervasive computing and ubiquitous computing, have considerable potential for enhancing the lives of many older people throughout the world and helping them to age in place (Sixsmith & Sixsmith, 2008). Ambient Assisted Living (AAL) refers to ICT systems, products, and services that integrate sensors, actuators, smart interfaces, artificial intelligence, and communications networks to provide more supportive environments for frail and disabled older people (Mokhtari, Khalil, Bauchet, Zhang, & Nugent, 2009; van den Broek, Cavallo, & Wehrmann, 2010). AAL has been an important emerging area of research over recent years involving collaboration between domain experts (health sciences, rehabilitation, gerontology, and social sciences) and technical experts (engineering, computing science, robotics). Research and development within AAL has aimed to develop applications and systems to facilitate independence (van den Broek et al., 2010), such as activity monitoring to detect potential emergencies, reminder devices for supporting and encouraging mobility and activities of daily living, monitoring activity patterns as indicators of change in cognitive and physical status, and smart interfaces to help people control their everyday environment. The European research project SOPRANO (Service-oriented Programmable Smart Environments for Older Europeans) developed supportive environments for older people based on the concept of AAL.

The SOPRANO system is an innovative integration of several new smart functionalities in the homes of older people. These functionalities allow the SOPRANO system to be used for various purposes in different situations. However, it is intended that users have a seamless experience in the sense that they perceive that they are supported by a single system. A major aim within the project has been to move away from technology-push and problem-focused approaches to user-driven approaches and how to explore, visualize, and map out an AAL system that will have practical benefits for users in their everyday lives.

In the final, field trials stage of the SOPRANO project, the subject of this chapter, the focus was on evaluating the possibilities of the integrated system and on demonstrating the spectrum of factors that influence the use of SOPRANO. The field trials consisted of two parts: *full function trials* and *large-scale field trials*. During the full function trials, the entire SOPRANO system was tested. Since the development was not completed for every single part of the SOPRANO system, feedback from users was gathered either as input to the design process and/or as evaluation on the

E. Avatangelou
Research Department, EXODUS S.A., 73–75 Mesogeion Ave. & Estias Str. 1,
115 26, Athens, Greece
e-mail: elav@exodussa.com

functionality of the total SOPRANO system. Since some innovative parts of the SOPRANO technology and services were still in the design process and were not yet proven to be reliable, to reduce participant burden, the full function trials were conducted in research facilities.

During the large-scale field trials, a subset of the SOPRANO system was installed in homes of end users. Different service levels were used to provide support to the users during the trials.

Acceptance is a key issue during field trials and is based on multiple factors that are described in the Smart Home Technology Acceptance Model (Bierhoff, 2011). It is important to differentiate between the acceptance of users exposed to the system for a short duration in a laboratory during the full function trials and the acceptance of users who experienced the system in their natural environment and daily life in the large-scale field trials.

This chapter starts with a description of user involvement, the SOPRANO system itself, and the use cases that were developed. This is followed by a description of the methodology and the framework used for analyzing the field trial results. In the next section a description of the field trial sites and the different stakeholders that participated in the field trials is given. This is followed by a description of the preparatory work for the field trials and the results of both the full function and the large-scale trials. The chapter concludes with an overview of lessons learned from the SOPRANO project.

8.2   SOPRANO User Involvement, Use Cases, and System

The SOPRANO project adopted a user-driven approach to ensure that technological development was usable, useful, and acceptable in the everyday life context of older users. Research, development, and design by, with, and for users were implemented at all stages of the product development life cycle. The focus on in situ use of SOPRANO facilitated the move from a health and social agenda that emphasized dependency to one that promotes active aging and creates supportive environments to enable healthy aging in the settings where older people live.

8.2.1   Iterative User Involvement

Users were actively involved at all four stages of the SOPRANO development process (see Fig. 8.1). The aim of the first stage, the elicitation of basic service requirements, was to find out which situations in daily life threaten the independence of older people or perceptibly limit their quality of life. Starting with an inventory drawn up by experts, situations that can threaten the independence of older people were collected. The inventory was complemented with information

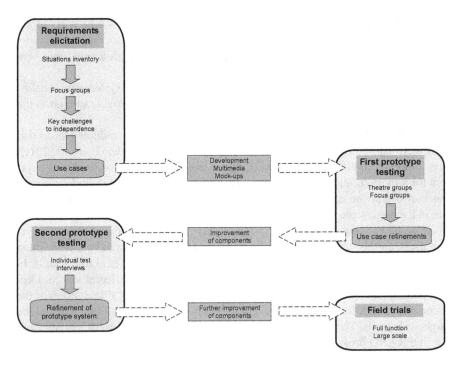

Fig. 8.1 Iterative development process in the SOPRANO project

from focus groups carried out with older people. Scenarios of new ICT-based services to address key challenges to independence were then developed. These included *use cases*—descriptive models—of how an AAL-based service can interact with a user in relation to a particular problem scenario (e.g., reminding a person about medication). Their purpose was to reflect not only the functionalities of the technical system under design but also the processes, actions, and interaction with the system of various actors/agencies: informal and formal carers, service provider agencies, GPs/hospitals, and the assisted older person her/himself.

Multimedia mock-ups and theater plays were developed to show the use cases to users during the first virtual prototype tests. Feedback from these was used to refine and improve the components in the use cases. The improved components were the subject of a second set of prototype tests that led to further improvement of the components. In the last stage were field trials where the system was installed and tested in laboratories (full function trials) and home settings (large-scale trials).

8.2.2   SOPRANO Use Cases

SOPRANO use cases do not merely reflect the development of a technical system and components but include a strong reference to the social world within which the person lives. This is crucial as the key to creating useful and acceptable technology-based services is to understand how they are embedded in a person's everyday life. The following use cases were used in the SOPRANO field trials (the type of trial in which each was used is shown in parentheses).

Medication reminding (full function and large-scale field trials): addressed the needs of older persons experiencing light to moderate forgetfulness who have to take medication at regular intervals and was targeted at improving adherence. For many older people, the ability to remain independent in their home depends on their ability to manage a complicated medication regimen (Dorman Marek & Antle, 2008). The full benefit of the many specific medications that are available can only be achieved if patients follow the prescribed treatment regimes reasonably closely (Osterberg & Blaschke, 2005). Poor adherence to medication regimes, particularly for older people, contributes to substantial worsening of quality of life, disease, death, and increased healthcare costs (Dorman Marek & Antle, 2008; Hayes et al., 2009; Osterberg & Blaschke, 2005). Reminders have been found to improve adherence (Hayes et al., 2009; World Health Organization, 2003). Older people can understand reminders spoken in a synthetic voice, provided the prompt texts are well designed, use familiar words, and incorporate redundancy (Wolters, Campbell, DePlacido, Lidell, & Owens, 2007).

> *At pre-set times (set by the person in charge of the medication regime) a reminder buzz will sound to alert the older person to take the medication from the dispenser. Additionally a reminder is either displayed on the TV (when turned on) accompanied with a synthetic auditory reminder. If the dispenser is not opened, repeated alerts will be activated by the system and after a certain period of time, an automatic message will be sent either to a professional carer or an informal carer.*

Remembering (full function and large-scale field trials): addressed users who experience increased forgetfulness in daily life situations, such as leaving the house without closing the windows or door, forgetting the house key or leaving appliances (TV, lights, etc.) running, and also forgetting important appointments. From research, it appears that older persons worry about their diminishing memory (Commissaris, Ponds, & Jolles, 1998), feel less in control of their memory functioning than younger people (Dixon & Hultsch, 1983), and are more upset when they forget run-of-the-mill events (Cavanaugh, Grady, & Perlmutter, 1983). A number of authors identify the loss of personal independence as a major concern for most elderly people (Bayer & Harper, 2000; Mynatt, Melenhorst, Fisk, & Rogers, 2004; Shafer, 2000; Starner, Auxier, Ashbrook, & Gandy, 2000). More and more AAL projects address these needs by designing applications that explicitly aim at increasing safety and well-being for elderly users (Röcker, Ziefle, & Holzinger, 2011).

> *When the user opens the front door to leave the house, SOPRANO checks the installed peripherals to determine if anything was forgotten to be switched off. If there was something still running, an alert message is displayed on the touch screen informing the user of the*

things that are not in order. The user can confirm that the warnings are read by touching the button 'I took care of it'. In case the user doesn't touch this button a message will be sent to an (in)formal carer. The user can also actively check the status of the house via the SOPRANO menu on the TV. Further to this, the user or a family member living in the household can enter important appointments into the system, via the user interface, and the user will be reminded of these appointments by SOPRANO.

Exercise facilitation (full function and large-scale field trials): focused on helping older people to improve their general fitness level by regular physical exercise. In order to improve cardiorespiratory and muscular fitness, bone, and functional health and to reduce the risk of noncommunicable diseases, depression, and cognitive decline, the World Health Organization (2010) recommends adults aged 65 years and above perform at least 150 min of moderate-intensity aerobic physical activity or 75 min of vigorous-intensity aerobic physical activity per week, or an equivalent combination of moderate- and vigorous-intensity activity. In SOPRANO an avatar was used to show how the exercises should be performed. Several recent studies (Kim, 2004; Ortiz et al., 2007) have concluded that conversational avatars are a good tool for obtaining a more natural communication with a user.

Whenever the exercises are due, SOPRANO will remind the user and, when s/he has given their consent, the exercises performed by an avatar will be displayed on the TV for the user to follow. The first time the exercises are performed, a carer/physiotherapist will be present to introduce the user to the new procedure.

Fall detection (full function trials): addressed older persons who are vulnerable and at risk of experiencing falls in the home. About 30 % of people over 65 years living in the community fall each year (Gillespie et al., 2008); 40 % of the people above the age of 75 need medical treatment after a fall (Davis, Robertson, Ashe, Liu-Ambrose, & Khan, 2010).

In case of a fall, the system will become active and first ask the user whether s/he is OK, and if the user confirms, the alarm is cancelled. If not, an alert message will be sent to the informal carer or professional carer (depending on the pre-set protocol) informing them about the situation and asking them to look after the user. During the whole procedure, SOPRANO will communicate to the user the steps it is taking to give them reassurance that help is on the way.

Entertaining (full function trials): addressed users who are becoming less and less active and are experiencing increasing boredom and loneliness. Loneliness and social isolation are often associated with older age and have been identified as risk factors for a number of health (both physical and mental) and related problems (Byles, Harris, Nair, & Butler, 1996; Grenade & Boldy, 2008; Savikko, Routasalo, Tilvis, Strandberg, & Pitkälä, 2005; Sugisawa, Liang & Liu, 1994; Walker & Beauchene, 1991). When providing interventions, it is important for service providers to acknowledge and accommodate individual differences (Findlay & Cartwright, 2002; Hicks, 2000).

The system is configured by the care provider to include a profile of the user's interests including leisure time activities, favorite TV shows, local community events, etc. The system is also configured to accommodate how often the user should be contacted, to ask how they feel and if they need any suggestions for activities for the day.

Activity monitoring (full function trials): addressed older people experiencing deterioration in their overall health status due to unfavorable changes in nutrition. The prevalence of malnutrition is relatively high in older people affecting over 10 % of the population aged 65 years and above (Baeyens, Elia, Greengross, & Rea, 1996).

> *The system supports the user in developing more favorable daily routines by reminding of meal times and suggesting recipes to be cooked.*

Safety monitoring (full function trials): addressed users experiencing difficulties related to sleeping who get out of bed and walk through the house at night, to make a cup of tea or to watch TV. For many, aging is associated with decreased total sleep time, increased sleep latency, and more awakenings particularly in the latter stages of the night (Baskett et al., 2001). In terms of sleeping disorders, the focus in SOPRANO was on sleep apnea.

> *When the user wakes up at night and gets out of bed, the SOPRANO system detects this via the bed occupancy sensor and will turn on the light beside the bed to prevent the user from stumbling and falling. After a predefined time of absence from the bed, the system will check if the user is OK, or otherwise, send an alarm to the carer. This check will be based on data from appliances (e.g. cooker on or off). Depending on the outcomes of the check, the system will contact the user to ask if everything is OK and send an alarm to the care provider if necessary. Sleep apnea can be detected without a person having to wear something on the body by a radar that is installed above the bed.*

8.2.3   SOPRANO Prototype System

For the use cases developed, a common hardware setup was created. This setup supported both the complete SOPRANO system shown in laboratories and the subset of the system installed in real homes.

In the labs, all use cases were installed and shown in detail to the participants. In general terms the use cases were divided into three categories:

1. Fully SOPRANO integrated (memory facilitating, fall detecting, medication reminding)
2. Content based (exercise facilitating, activity monitoring, entertaining)
3. Very innovative (safety monitoring)

The home installations were comprised of a subset of the SOPRANO system: a selection of use cases that were selected based on security and reliability, user-system interaction, inclusion of carers, frequency of need, and the development of devices. The system subset supported the following use cases: medication reminding, memory facilitating, and exercise facilitating. A selection of the equipment used for the in-home large-scale field trials is presented in Fig. 8.2.

Besides the equipment displayed in Fig. 8.2, a SOPRANO PC was supplied that ran the services. Other additional devices were the TAPIT module to connect the Connect+ to the SOPRANO PC, a video cable to connect the SOPRANO PC with the TV, a TV tuner/frame grabber to feed the TV signal to the iTV module running

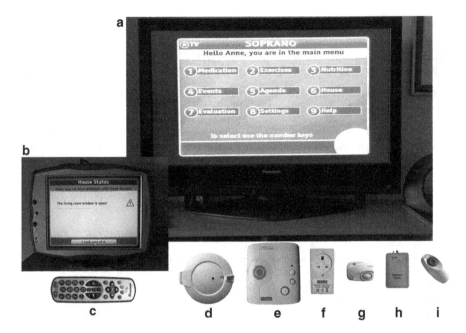

Fig. 8.2 Selection of equipment used for large-scale field trials. (a) SOPRANO main menu available on the TV. (b) Touchscreen installed next to the main entrance displaying warning messages when leaving the home in case home appliances are still running or windows are open. (c) Remote control used to operate the SOPRANO menu on the TV. (d) Medication dispenser. (e) Home telecare platform (Connect+) that receives messages from medication dispenser, electrical usage sensor, fall detector, universal sensor and security red button. (f) Electrical usage sensor to monitor the electrical consumption of home appliances. (g) Fall detector. (h) Universal sensor to detect if a window or door is open. (i) Security red button

on the SOPRANO PC, and a remote control interface that received input from the SOPRANO remote control.

8.3   Methodology for Field Trials

To decide which methods of evaluation to use in the full function trials and the large-scale field trials, the well-established theory of care quality by Donabedian (1969) and its extension into three quality perspectives by Øvretveit (1992) offered a fruitful approach. This model of evaluation defines a service in terms of inputs, process, and outcomes to be evaluated from the perspectives of the clients (patients), professionals, and managers. These three perspectives refer to the three central stakeholders in care provision and allow quality to be evaluated from three perspectives:

Table 8.1 SOPRANO multi-method approach to user-driven research

Stakeholder/ perspective	Inputs	Process	Outcomes
Older person (client, patient)	*Personal characteristics*: Psychological Physical Cognitive	*Acceptability*: Is the assisted person prepared to accept the system in everyday living context and what features are preferred or disliked?	Well-being Perceived independence Remain living at home Increased participation
	Contextual factors: Home environment Neighborhood	*Usability*: How user friendly is the system, what barriers exist in its use, how well does it fit the social and living context of the person?	
Informal carer	Support network	Involved in some SOPRANO use cases. Also sees the effects on the older person	Reduced anxiety Reduced stress
Care professional (formal carer)	Care services	Involved in some SOPRANO use cases. Also sees the effects on the older person	Improved quality of service
Care management	Financial resources	Integration of SOPRANO use cases in care provision	Reduced costs, e.g., reduction in hospitalization
Building/technical	Buildings	Integration of SOPRANO system in new and existing buildings	Improved market value of property

1. *Patient quality*: the service provides patients with what they want and expect, during and after the service.
2. *Professional quality*: the service follows procedures and methods which are thought to be most effective in meeting patients' clinical needs, as assessed by health professionals.
3. *Management quality*: the service uses available resources in the best way to achieve patient and professional quality, without waste and within higher level requirements.

This model was used to frame the evaluation of the SOPRANO system. Although SOPRANO did not focus on clinical care, the service-oriented approach of the model was still applicable. However, due to the nature of service delivery in the home environment, additions were made to the list of perspectives. The perspective of the informal carer was added because in the home environment, care provision is often divided between self-care, informal care, and formal care. Furthermore, since the SOPRANO system needs to be installed in the home, the perspective of building stakeholders, for example, a housing association or a property developer, is also relevant. The chosen multi-method approach is displayed in Table 8.1.

8.3.1   Full Function Trials Methodology

SOPRANO aimed to find out how users accepted a very complex, integrated system. To do so, in the full function trials, users were invited to laboratories where they could experience the whole integrated system and give feedback. In the laboratories, use cases were used to explore the possibilities and show the added value of the SOPRANO system. By doing so, it was also possible to show the connectivity between different parts of the system. Participants in the full function trials took part in a guided exploration of the SOPRANO system, with some limited possibilities for free exploration, or took part in an in-depth analysis of specific parts of the SOPRANO system.

The full function trials started with a pretest to evaluate (1) the equipment and use cases, (2) protocols, and (3) the questionnaires. Testing the equipment and use cases explored the integration of the components and the manner in which the use cases could be presented. Besides the technical integration, it was also important to decide how the interconnectivity between all components could best be shown to future participants. Pretesting of the protocols and the questionnaires was completed in two steps. First, the plans for the guided tours and evaluation materials were assessed by experts and their comments were taken into account. Second, before the guided tours commenced, a group of older persons was invited to give their comments on a preliminary version of the guided tour.

The participants for the subsequent guided tours included representatives from the five stakeholder groups described in Table 8.1. The majority were older persons who would be the main users of the SOPRANO system. The inclusion of stakeholders who do not directly interact with the system, but who play a role in service delivery, was a significant departure from previous research cycles.

In a second part of the full function trials, an in-depth analysis was conducted of the graphical user interface (GUI) on TV and touch screen. The main aim was to test how alterations, based upon improvements suggested by participants in previous research cycles, were perceived. Additionally, some tests were performed on the Admin GUI, which provided informal and formal carers with the possibility of uploading information to the SOPRANO system.

8.3.2   Large-Scale Trials Methodology

To find out how users accepted the system at home in their own environment, user homes were equipped with a subset of the SOPRANO system. To prepare for the in-home trials, each site set up a test environment. The purpose was twofold. First, the trial site would gain experience in setting up the system and could check if the system met their own standards for installing technology in homes of service users. Second, when the system was running, potential participants could be invited to see it before they made a final decision about whether or not to participate in the large-scale trial.

The large-scale trials involved older persons, informal carers, care organizations (focus on user and technical aspects), and local installers. They each played a specific role in the implementation process of the SOPRANO use cases in real homes. The following are the steps in the process and the participants involved (in parentheses):

- Setting up the test environment (care organization): This step involved the setup of a test environment at each large-scale trial site.
- Visit to test environment (care organization, older persons, informal carers): Users might have difficulties in imagining how the system would work. Therefore, it was decided to invite them to the test environment and show them the system to help them with their decision to participate in the trial.
- Technical checklist (care organization, local installer): This included a technical assessment of the home, covering aspects such as Internet connection, TV set, TV signal, and telephone line.
- Acquisition questionnaire (older person, informal carer): These questionnaires contained the inclusion/exclusion criteria, and were used during the recruitment process, in order to derive an appropriate sample of persons that could participate in the large-scale field trials.
- Installation visit (care organization, local installer): In cooperation with the user, carer, and local installer, a date was set when the SOPRANO system would be installed.
- Preinstallation interview (older person, informal carer).
- Post-installation interview (older person, informal carer).

The preinstallation interview was an in-depth individual interview focused on the attitudes, expectations, and technical experience of users and their informal caregivers. The post-installation interview focused on users' and informal caregivers' overall impressions of the SOPRANO system, expectations, attitudes, and feelings towards the system, perceived benefits and added value, and utility of the SOPRANO system. Older persons reported upon their own experience trying out the SOPRANO system. Informal carers commented on how they interacted with the SOPRANO system and about the way the older person they cared for felt about the SORPANO system, the changes it had made in the daily life of the older person, and whether or not the older person was overstrained by the system.

8.4 Framework for Analyzing Results from Field Trials

During the field trials, a variety of responses from participants were collected. The Smart Home Technology Acceptance Model presented below was used as the framework for analyzing the information collected. A mix of quantitative and qualitative methods was used to evaluate these trials.

8.4.1 Smart Home Technology Acceptance Model

The Smart Home Technology Acceptance Model (SHTAM) shown in Fig. 8.3 is based on three models. The first, the *USE-Model* (Dewsbury, Sommerville, Rouncefield, & Clarke, 2002), considers the concept of home as an interrelationship among three overlapping spheres—the user, the system, and the environment. The second was a model of the acceptability of assistive systems developed by McCreadie and Tinker (2005), the most important aspect of which is what the authors call *felt need for assistance*. Felt need depends on several user characteristics, the housing type and design, and the interaction of these variables. The third model was a reformulation of the Technology Acceptance Model (TAM) (Davis, 1989) that takes into account other theories about technology and behavior adoption as the Theory of Planned Behavior and Uses and Gratifications Theory (Venkatesh, Morris, Davis, & Davis, 2003). TAM suggests two factors as the principle antecedents of attitude towards a system: perceived usefulness and perceived ease of use. Many studies (Gardner & Amoroso, 2004; Lu, Yu, Liu & Yao, 2003; Nyseveen,

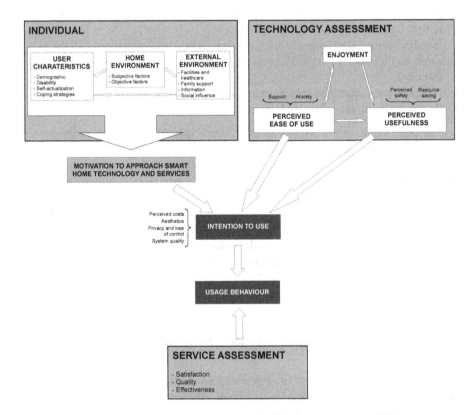

Fig. 8.3 Smart Home Technology Acceptance Model (Bierhoff, 2011; Conci, 2008)

Pedersen, & Thorbjørnsen, 2005) confirm that a more situation-specific model needs to be added to the TAM.

The SHTAM model is divided into three parts: *individual, technology,* and *service* related. The individual part of the model combines the characteristics of the user, the home environment, and the external environment. User characteristics include demographic information and also information about the psychological and physical condition of the person. The home environment part of the model includes subjective factors related to the perception of safety and comfort in the home and objective factors related to accessibility. The external environment part concerns facilities, services, family, and other sources of support and social influence.

The conjunction level, between the individual and the intention to use smart home technology and services, is called *motivation to approach smart home technology.* It is a transaction state that could be compared to the felt need for assistance of McCreadie and Tinker (2005). The characteristics of the user, the home environment, and the external environment together influence the motivation of the user to approach smart home technology and services. Motivation is related to expectations, attitudes, and feelings towards the system, in this case, the SOPRANO system, before actually using it.

The technology part of the model refers to the actual use of the system and its functionalities. The basis for this part of the model is the TAM with the two main factors: perceived usefulness and perceived ease of use (Davis, 1989). A third factor is added, namely enjoyment, which refers to the extent to which the activity of using the system is perceived to be enjoyable in its own right, apart from any performance consequences that may be anticipated (Davis, Bagozzi, & Warshaw, 1992). Enjoyment is influenced by ease of use and will influence perceived usefulness (Kwon & Chidambaram, 2000; Yi & Hwang, 2003). Perceived ease of use is defined as the degree to which a person believes that using a system will be free from effort (Davis, 1989) and is influenced by support (Phang et al., 2006; Venkatesh & Morris, 2000) and anxiety (Venkatesh & Morris, 2000). Perceived ease of use influences perceived usefulness as an intention to use the system (Legris, Ingham, & Collerette, 2003). Perceived usefulness in this model consists of the benefit and added value of the system and is influenced by perceived safety against unexpected events and resource saving related to cost and time (Phang et al., 2006). A significant body of TAM research has shown that perceived usefulness is a strong determinant of intention to use a system (Venkatesh & Morris, 2000).

Intention to use the system is influenced by the motivation to approach the smart home technology and services (Gardner & Amoroso, 2004; Lu et al., 2003; Nyseveen et al., 2005), perceived costs (Kleijnen, Wetzels, & de Ruyter, 2004), aesthetics (Veryzer, 1995), privacy and loss of control (Wilkowska & Ziefle, 2011), and system quality (Kleijnen et al., 2004). System quality is composed of reliability, flexibility, integration, accessibility, and timeliness (Wixom & Todd, 2005).

Intention to use the system is reported to have a significant relationship with actual usage (Legris et al., 2003). The actual use of the system also depends on the whole chain of steps that occur behind it. It is not possible to evaluate a technological system without also taking into account the network of people that need to offer assistance. In other words, evaluating a smart home system requires an assessment of all other factors that it is related to, i.e., service providers and provided service. In the model, the assessment of service provision is influenced by satisfaction, quality, and effectiveness.

8.4.2 Data Collected During Field Trials

During the full function trials, every participant filled in an acquisition questionnaire before visiting the laboratory. The questions included in it were based upon the individual-related part of the Smart Home Technology Acceptance Model. During the introduction of SOPRANO, which was the starting point of the guided tour, the project and the use cases were explained in detail. Participants asked clarifying questions and commented on the use cases. After the demonstration of a specific use case, each participant filled in an assessment questionnaire covering the technology and service assessment part of the model as well as intention to use. During the demonstration of the use cases, much attention was devoted to discussing the perceived usefulness, perceived ease of use, intention to use, and service provision. Minutes were taken of the discussions. During the usability tests, data were collected by making notes of comments and observing participants' actions when they performed tasks while thinking out loud during a detailed walk through.

Additionally, during the large-scale trials, every participant filled in an acquisition questionnaire before the subset of the SOPRANO system was installed in their home. Information regarding their technical experience, their expectations, and their attitudes towards the SOPRANO system was collected during the pre-interview with older persons and informal carers. The post-interview examined how the older person and the informal carer used SOPRANO and how it impacted on their lives. Besides these formal procedures, the user and technical experts from each trial site made notes of their own experiences related to their own role in the implementation process of the SOPRANO use cases in real homes.

8.5 Field Trial Sites and Participants

This section describes the locations for the field trial research and the participants who took part in the field trials.

8.5.1 Field Trial Sites

The full function (laboratory) trials were conducted in three different geographic locations in Europe. Consortium partners of SOPRANO were responsible for the lab at each site. Smart Homes was the consortium partner responsible for the lab in

Eindhoven, Netherlands; INGEMA in San Sebastian, Spain; and the London Borough of Newham for the lab in Newham, UK. For the large-scale trials (home installation), there were four sites. Archipel was the consortium partner responsible for the homes in Eindhoven, Netherlands; Fundación Andaluza de Servicios Sociales (FASS) for the homes in Malaga, Spain; the West Lothian Council for the home in West Lothian, UK; and the London Borough of Newham for the home in Newham, UK. The following figures provide an impression of activities at the different field trial sites (Figs. 8.4, 8.5, and 8.6).

8.5.2 Field Trial Participants

Participants who took part in the field trials for SOPRANO were experts from the consortium, local installers, older persons, informal carers, formal carers, and other stakeholders. Involving a broad range of stakeholders was seen as important for establishing the impact of SOPRANO from the multiple perspectives of those involved in delivering care.

A distinction can be made between experts from the consortium who took part in the user-related research and experts who were involved as technical experts. Professionals who work in the field of assistive technologies and user-system interaction experts were involved in assessing the integrated SOPRANO system. They were responsible for the final check of the SOPRANO system before other stakeholders were invited to take part in the field trials.

Experts in user-related research were an interdisciplinary team of researchers consisting of user-system interaction specialists, psychologists, and gerontologists. Their role was to fine-tune the way the use cases would be shown and implemented and to have a detailed look as to how the suggested methods would work out in practice with the current status of the SOPRANO system.

To set up the full function sites as well as the test environments at the large-scale sites, a local technical expert was engaged. The technical expert was in close contact with SOPRANO technical partners when setting up and testing the SOPRANO system. Problems related to the functioning of the use cases and practical installation issues were detected and solved.

In preparation for the installation in older persons' homes, the local technical expert visited each property to assess if the home was suitable to take part in the SOPRANO research and to make sure that all technical preconditions were met before the actual installation. Local installers were responsible for equipment installation and ongoing support during the trial period.

The other stakeholder groups who participated in the field trials represent the multi-method approach to user-driven research as presented in Table 8.1 and are described below.

A total of 189 older persons took part in the SOPRANO field trials; their mean age was 74. When asked if they experienced any aging-related problems, 31 % indicated that they did not. Among the participants who did face problems, most were

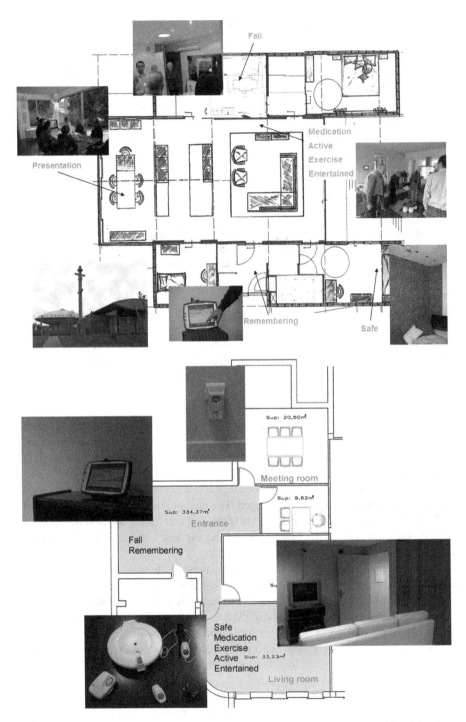

Fig. 8.4 SOPRANO full function trials in Eindhoven, Netherlands (*top photo*), and San Sebastian, Spain (*bottom photo*)

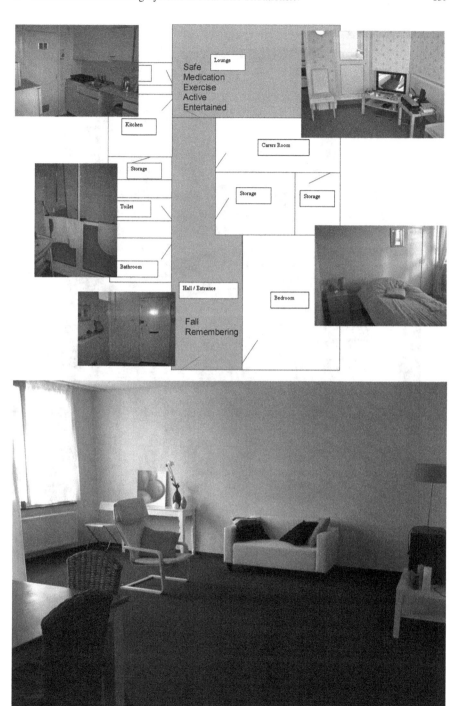

Fig. 8.5 SOPRANO full function trials in Newham (*top photo*) and large-scale test environment in Eindhoven (*bottom photo*)

Fig. 8.6 SOPRANO large-scale research in Malaga (*top photo*) and West Lothian (*bottom photo*)

related to declining physical health and mobility, fear of crime, forgetfulness, and financial difficulties (see Fig. 8.7).

Only a few people took no medication ($n = 19$). The majority of participants (45 %) took more than four types of medication (Fig. 8.8). Participants were aware of the reasons that they needed to take medication. Most health problems

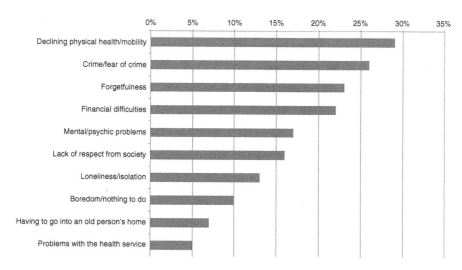

Fig. 8.7 Problems experienced by older persons ($n = 137$, use cases in general)

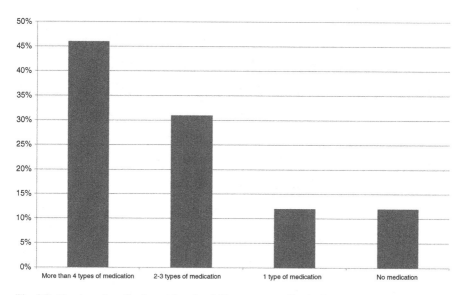

Fig. 8.8 Number of medications taken ($n = 166$, use case medication)

reported were related to high blood pressure, high cholesterol, heart rhythm disorder, and diabetes.

Regarding self-reported forgetfulness (Fig. 8.9), categories were as follows: $1 =$ never or a few times per year, $2 =$ a few times per month, $3 =$ a few times per week, and $4 =$ every day. The number of participants who reported some form of

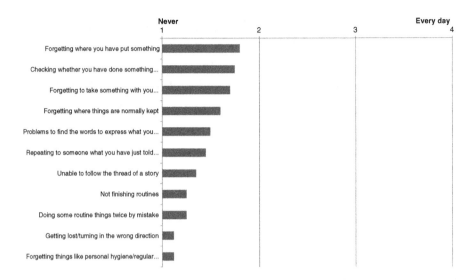

Fig. 8.9 Forgetfulness among the older people (self-assessment, $n = 131$, use case = remembering)

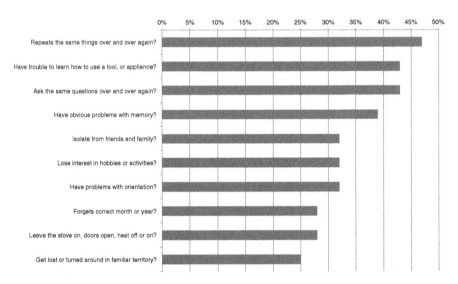

Fig. 8.10 Forgetfulness among the older people (assessed by their informal carers, $n = 28$, use case remembering)

memory problem was very low. Among those who did report problems, the most common were "forgetting where you have put something" (mean = 1.81; s.d. = 0.88) and "checking whether you have done something that you meant to do" (mean = 1.74; s.d. = 0.69).

In addition to the older persons' self-assessment of forgetfulness, the informal carers were asked to provide a more "objective" external judgment. According to the informal carers (see Fig. 8.10), the assisted persons most often tended to repeat the same things (46 %) and questions (43 %) over and over again. It is also worth mentioning that 43 % of the carers had trouble learning how to use a tool or appliance.

The average age of the 28 informal carers who participated was 55 years. In relation to the relatives whom they looked after, a significant number of informal carers either lived in the same home ($n = 10$) or resided less than 5 km away ($n = 11$). Most informal carers visited and/or called their relative daily or at least several times per week. The majority of informal carers (63 %) reported that they "do not feel strained" in caring "for and about" their relative, whilst 15 % indicated that they felt very strained by caring for and about their relative.

The group of formal carers ($n = 24$) who took part in the full function guided tours represented a diverse group. In terms of expertise, the following jobs/disciplines were represented: care and service consultants, case managers, specialists in gerontological nursing, psychologists, geriatric physiotherapist, and occupational therapists.

The care management group ($n = 11$) consisted of representatives from care provider organizations, decision-makers responsible for establishing care strategies, and those responsible for administering and managing budgets. The building/technology group ($n = 10$) consisted of representatives of installation companies, housing organizations, property developers, builders, and suppliers of smart home technology.

8.6 Field Trial Results

This section highlights the key findings of the SOPRANO field trials. It starts with a description of the delicate process of moving the SOPRANO system from a laboratory environment into real homes. The results of the field trials are then presented using the Smart Home Acceptance Model as the conceptual framework. When analyzing the results, an equal amount of attention was paid to similarities as well as to differences between the trial sites, since differences are equally important when it comes to implementing technology in real life.

8.6.1 From Laboratory to Real Home

The process of moving technological innovations out of the laboratory into real homes is very delicate and must not be underestimated as the following examples highlight.

Several pretests were planned for the full function trials. Testing the equipment and use cases focused on the integration of the components in the demonstration home and the way the use cases could be presented. Specific attention was paid to the integration process of the available technology. Results of this test were valuable

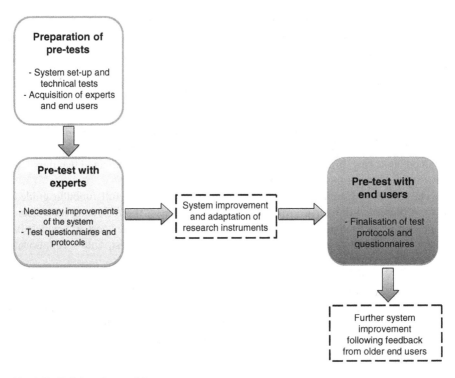

Fig. 8.11 Full function workflow pretest

not only for the full function trials but also for the large-scale trials. Besides the technical integration, it was also important to decide how the interconnectivity between all components could be shown in the best way to future participants. With respect to the actual conduction of the research, the pretest focused on the test protocols and the questionnaires. The workflow for the full function pretests is shown in Fig. 8.11.

During the integration process, it became apparent that components communicated with each other but the fine-tuning was missing. Some of the actions required were just simple tests to check for inconsistencies and to make sure that those details were covered in the technical components; others were more serious and related to the stability of the system. Initially it was anticipated that issues and problems with the SOPRANO system would be identified and overcome during the full function trials. In practice, while some issues concerning the integration of the components and stability of the system were resolved, others, mainly related to the stability of the system, remained. During the full function research, integration issues could be overcome by the use of simulating techniques, and stability issues had a minor impact due to the fact that a moderator showed the use of the use case. However, these solutions could not be used in the large-scale trials where the SOPRANO needed to run as a stable system 24/7 without any interference by a moderator.

Each large-scale trial site set up a test apartment to gain experience with setting up the subset of the system and to check if it met their own standards for installing technology in homes of service users. After this check, potential participants were invited to visit the test apartment. Older persons were very engaged and curious during the visit and reported having a greater sense of confidence in being able to use the equipment after the visit. Informal carers felt that the visit gave them a deeper understanding of how the system worked and how it could support their loved ones.

The work of the local technical experts was vital to the success of the field trials. They were the local contact point for participants and care personnel, solved minor issues, and cooperated with SOPRANO technical experts in case of major issues.

With respect to the actual installation, it became clear that each home was different. All sites reported that in most cases the technical checklist was sufficient to meet the preconditions but it was not detailed enough to cover all the differences between individual homes.

8.6.2 Individual

The characteristics of the older persons who participated in the field trials, and comments made by other stakeholder groups, indicated that the SOPRANO system would be of most benefit to older people who were in the early stages of decline, yet retained medium to high levels of mobility and independence, although requiring some level of care support.

Older persons saw their home as an important place, evoking a strong sense of identity through feelings of attachment, familiarity, and belonging. Home was somewhere they wanted to remain and evoked happy memories. Therefore, it was very important that the system did not take over the environment and that the characteristics of the home remained as they were before the installation. People were also worried about assumptions that visitors might make upon seeing the system, creating a stigma that they were in need of care.

A general comment was that buildings are often not suitable for the current demands of their residents. In relation to the fall use case, formal carers commented about the importance of removing loose carpets and about placement of furniture, and wires, in preventing falls.

Older people had the feeling that technology could help support their independence: "*I am always relying on others and my family. Technology means I can move from dependency to independency.*" Informal carers living at a distance indicated that technology could give them peace of mind when they are not able to visit. However, family members also valued the role of service centers and care organizations in ensuring that there was assistance and intervention when required.

Figure 8.12 shows the support older persons currently received. They expected the system to help them to live longer into old age through supporting them to live independently within their own homes. They had high levels of motivation for

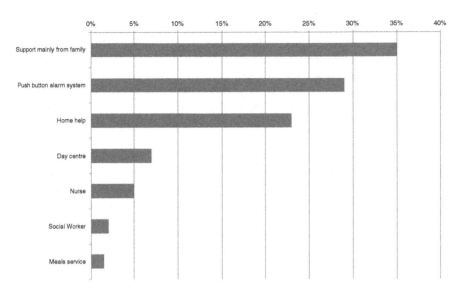

Fig. 8.12 Services received by the older persons (*n* = 137)

involvement in the trial. This originated from a feeling that they were being involved in something that would benefit them and other older people in the future. They envisioned the system to be easy to use and responsive even in the event of breakdown and expressed the need to have access to a local contact should the system fail. Informal carers reported high levels of motivation arising from the potential that the technology would provide an extra layer of support for the people they were caring for.

8.6.3 Technology Assessment

With respect to enjoyment, all participants showed enthusiasm for being involved in the trials and in trying something new. Furthermore, they responded enthusiastically to features of use cases they found appealing. For example, an older person responded to the exercise use case: *"It's like I've got a personal trainer in my home!"*

In terms of the perceived ease of use of the interfaces, the usability tests and actual use in real homes indicated that the participants had a positive perception regarding the ease of navigation through the menus on the TV and touch screen, the use of the remote control, and comprehensibility of the icons and text on the displays. Difficulties were experienced by informal carers when filling and programming the pill dispenser and with manually putting data into the administrator module. Additional support was needed for performing those two tasks. Related to the content of the use cases, some participants clearly described the interconnectivity between the different system components which promoted ease of use and prevented unnecessary alerts. For example, *"If the medication is taken you won't get the notification on TV, so they communicate. I think that is good."* Also recommendations were made to improve ease of use: *"The instructions aren't clear enough for some of the specific exercises. This can be dangerous and an older person could fall."* Participants did not initially feel anxious about using the SOPRANO system. However, increased levels of anxiety were reported due to the false alarms, system malfunction, and interference with the Internet and television.

While the general opinion was that this was a good system that satisfied older persons' needs, during the large-scale field trials, many users found it difficult to evaluate the perceived usefulness of the SOPRANO system because the system was unstable and there were frequent crashes, while other functions did not work at all. However, users felt that there was significant potential in the system in supporting them to live independently when it was functioning efficiently. A striking difference was noted in the fall detection use case. A person who has fallen once commented: *"From previous experience this seems pleasant."* A person who had not fallen commented: *"Very useful if you are ready for it."* It seems, in other words, that people have to cross a certain threshold to admit that they need the fall detection subsystem. On the other hand, the SOPRANO system was seen as having potential to increase the security and safety of older people and reduce property crime: *"I thought I had closed the windows. However the SOPRANO system triggered an alarm when I left home, making me come back and close the window."*

8.6.4   Service Assessment

A major issue about service assessment was related to the question *"who is responsible for what?"* The following example highlights some of these concerns in relation to the medication use case. Experts, (in)formal carers, and older persons raised the question of who would be responsible for filling the pillbox. Care management commented that it would be: *"Difficult from a practical point of view. Who will take care of filling the pill dispenser? Home care employees don't have permission to do it."* Older persons commented: *"The person who fills this must know what is needed."* The medication use case also needed to be responsive to changes in the medication plan. Weekly changes were often made to prescriptions and these would need to be reflected in the medication use case. One participant commented: *"what in case of sudden changes to the medication plan?"*

A second point raised was that the focus should be on integrating SOPRANO services in the entire palette of service provision towards an older person. Participants felt that it was currently difficult to align the work of general practitioners, healthcare professionals, and formal carers and that the introduction of the SOPRANO system added another layer of complexity.

Thirdly, participants felt that there was a need to ensure that a procedure was in place in the case of a system breakdown, whereby technical support would visit to repair any problems. There was a need for this service to be responsive, to ensure that the safety of the older person was not compromised.

8.6.5 Intention to Use and Usage Behavior

A key aspect mentioned in relation to the intention to use the system was the notion of personalization. Older persons felt that the SOPRANO system needed to be tailored to the individual needs of the user. This would ensure that redundant aspects of the system were removed if they were not needed and that these elements could be added if the needs of the older person changed.

Participants were concerned about who would bear the responsibility for the costs of the system. Costs included ongoing energy costs, hardware costs, maintenance, and updates. It was difficult for older people to determine if the costs outweighed the benefits, especially for those who reported higher levels of mobility and positive health: *"You don't know at this moment if it's worth the investment."* Care management and representatives from the building and technology sector stated that they were convinced that by integrating these kinds of systems in buildings and care provision, money could be saved on installations, costs of buildings, and costs of care provision. Dependency on technology was an issue raised by participants. There was concern that the system may suffer a malfunction and that the older person would be endangered if they were dependent on it for daily functioning: *"You get dependent from all of this, for instance what happens in case of an electricity failure?"* In the large-scale trials, participants had initial concerns that the system might be "doing the thinking" for them and that it would replace independent decision-making. However, when engaging in the trial, older persons indicated that they did not see the reminders as losing a sense of control; rather they welcomed the suggestion to do exercise, thereby encouraging independent action, as opposed to dependency. Informal carers felt that future technologies might help when caring for older people, enabling them to remain at home longer. However, they stressed that: *"Technology shouldn't replace people. People need people."*

All participants had the intention to use the system, but actual use varied across participants. High levels of acceptability were reported because they could access it at any time of the day. The flexibility of having a system within the home environment, which they could use in and around their everyday commitments, was welcomed. User ratings of whether they would use the SOPRANO system in the future were affected by the ongoing instability of the system. For example, an older person at one of the large-scale trial sites commented: *"If reliability issues are resolved I would consider having it long-term."*

Figure 8.13 shows some of the participants taking part in the field trials of SOPRANO.

8.7 Lessons Learned from the Field Trials

All stakeholders who took part in the research demonstrated a willingness to be involved in the application and development of the SOPRANO system. They demonstrated high levels of motivation when using the technology and being

Fig. 8.13 SOPRANO field trial participants, using an exercise avatar (*top photo*) and touch screen technology (*bottom photo*)

engaged in the research. This originated from an enthusiasm to assist in research that might potentially benefit older persons to live independently in their own homes.

Older persons felt that an increased sense of independence resulted from the ability to undertake activities of daily living and technology was perceived as an intervention to assist them in that. Yet, older people revealed a dilemma in retaining a sense of independence while realizing the need for intervention and care. Independence and dependence are not fixed and often had to be negotiated as the needs of the older person change. This is closely linked with the issue of personalization to ensure that services are suitable to the unique and changing needs of the older person. Furthermore, a flexible approach to care is needed, making it possible to easily switch between levels—self-support, assistance of an informal carer, and assistance of a formal carer. Ongoing issues with the reliability of the SOPRANO system prevented the concept of independence and personalization from being explored further. However, based on these observations, further work on the influence of the amount of personalization and adaptability on perceived usefulness is expected. This is strengthened by the fact that most of the participants indicated that they were not ready to use a system like SOPRANO yet, as it provided more support than needed at the moment.

Another variable in the Smart Home Acceptance Model that is in need of further investigation is aesthetics. Several comments, especially during the large-scale trials, were made concerning the stigma associated with assistive technology and the negative assumptions visitors might have when seeing the system.

Participants identified the potential for SOPRANO to bring about changes in existing modes of care delivery. Older people felt that such a system could enable them to be more independent, alleviating the care burden upon informal carers. However, they did not want this to result in less face-to-face contact with informal carers, as this was an important source of social support. Informal carers reported that assistive technology would help alleviate the worry and stress of caring for older people, providing an extra layer of safety and security. The technology also had the potential to increase the quality of time that informal carers spent with older people as it reduced the burden upon them to be caring for the older person.

In relation to service delivery, additional attention is needed to align the responsibilities of all involved in care delivery to an older person. A precondition for this is that a focus on integrated care should include technological possibilities. Procedures should also be in place for maintenance and problem handling.

8.8 Conclusion

The focus of the SOPRANO project was on promoting active aging and health by moving away from dependence. Participants indeed reported that a greater sense of independence for older persons could be achieved when using SOPRANO. However, many felt that they did not need the system right now.

By starting the user-centered design process and looking at key challenges to independent living, the approach was very much problem oriented. An addition to that approach could be one that focuses more on things that make people happy and are considered to be leisure and fun.

There is often the assumption that a system operating effectively within a lab environment will do so within the home, yet there are a number of situational and contextual factors which impact on system reliability. A simplified and evolutionary approach to trialing new technology would be one that allows participants to learn and use individual components, progressing to an advanced stage only when the older person feels comfortable and when the operability of the system can be guaranteed. This gradual installation process could start with leisure-oriented services.

With respect to adoption of the developed technology and service after the conclusion of the research project, care organizations should consider offering services not just because the client is in need of care but because the client is willing to pay for additional comfort.

Acknowledgements SOPRANO (http://www.soprano-ip.org/) is an Integrated Project funded under the EU's FP6 IST program Thematic Priority: 6.2.2: Ambient Assisted Living for the Aging Society (IST—2006—045212). The authors acknowledge the input and role of the SOPRANO consortium and would also like to thank the many people who volunteered in the various stages of user research described in this chapter.

References

Baeyens, J. P., Elia, M., Greengross, S., & Rea, N. (1996). *Malnutrition among older people in the community: Policy recommendations for change*. London: European Nutrition for Health Alliance.

Baskett, J., Wood, C., Broad, J., Duncan, J., English, J., & Arendt, J. (2001). Melatonin in older people with age-related sleep maintenance problems: A comparison with age matched normal sleepers. *Sleep, 24*(4), 418–424.

Bayer, A.-H., & Harper, L. (2000). *Fixing to stay: A national survey on housing and home modification issues – executive summary*. Washington, DC: American Association of Retired Persons.

Bierhoff, I. (2011). *Adapted version of the smart home technology acceptance model* (Internal report). Eindhoven: Smart Homes (Initial model described by Conci and Bierhoff in 2008).

Byles, J., Harris, M., Nair, B., & Butler, J. (1996). Preventive health programs for older Australians. *Australian Health Promotion Association, 6*, 37–43.

Cavanaugh, J., Grady, J. G., & Perlmutter, M. P. (1983). Forgetting and use of memory aids in 20 to 70 years olds everyday life. *International Journal of Aging and Human Development, 17*(2), 113–122.

Commissaris, C., Ponds, R., & Jolles, J. (1998). Subjective forgetfulness in a normal Dutch population: Possibilities for health education and other interventions. *Patient Education and Counseling, 34*(1998), 25–32.

Conci, M. (2008, internal report). *Initial smart home technology acceptance model*. Eindhoven: Smart Homes.

Davis, F. (1989). Perceived usefulness, perceived ease of use and user acceptance of information technology. *MIS Quarterly, 13*(3), 319–340.

Davis, F., Bagozzi, R., & Warshaw, P. (1992). Extrinsic and intrinsic motivation to use computers in the workplace. *Journal of Applied Social Psychology, 22*, 1111–1132.

Davis, J., Robertson, M., Ashe, M., Liu-Ambrose, T., & Khan, K. (2010). International comparison of cost of falls in older adults living in the community: A systematic review. *Osteoporosis International, 21*(8), 1295–1306.

Dewsbury, G., Sommerville, I., Rouncefield, M., & Clarke, K. (2002). *Bringing IT into the home: A landscape documentary of assistive technology, smart homes, telecare and telemedicine in the home in relation to dependability and ubiquitous computing* (DIRC working paper PA7 1.1).

Dixon, R., & Hultsch, D. F. (1983). Metamemory and memory for text in adulthood: A cross-validation study. *Journal of Gerontology, 38*, 689–694.

Donabedian, A. (1969). Some issues in evaluating the quality of nursing care. *American Journal of Public Health, 59*, 1833–1836.

Dorman Marek, K., & Antle, L. (2008). Medication management of the community-dwelling older adult. In R. G. Hughes (Ed.), *Patient safety and quality: An evidence-based handbook for nurses*. Rockville, MD: Agency for Healthcare Research and Quality (US).

Findlay, R., & Cartwright, C. (2002). *Social isolation and older people: A literature review*. Brisbane: Australasian Centre on Ageing, The University of Queensland.

Gardner, C., & Amoroso, D. L. (2004). *Development of an instrument to measure the acceptance of internet technology by consumers*. Paper presented in Proceedings of the 37th Hawaii International Conference on System Sciences, San Diego State University, CA.

Gillespie, L., Gillespie, W., Robertson, M., Lamb, S., Cumming, R., & Rowe, B. (2008). Interventions for preventing falls in elderly people (review). *The Cochrane Library 2008*, (4).

Grenade, L., & Boldy, D. (2008). Social isolation and loneliness among older people: Issues and future challenges in community and residential settings. *Australian Health Review, 32*(3), 468–478.

Hayes, T., Cobbinah, K., Dishongh, T., Kaye, J., Kimel, J., Labhard, M., et al. (2009). A study of medication-taking and unobtrusive, intelligent reminding. *Telemedicine and e-Health, 15*(8), 770–776.

Hicks, T. (2000). What is your life like now? Loneliness and elderly individuals residing in nursing homes. *Journal of Gerontological Nursing, 26*, 15–19.

Kim, Y. (2004). *Pedagogical agents as learning companions: The effects of agent affect and gender on student learning, interest, self-efficacy, and agent persona* (Ph.D. Thesis). Tallahassee, FL, USA.

Kleijnen, M., Wetzels, M., & de Ruyter, K. (2004). Consumer acceptance of wireless finance. *Journal of Financial Services Marketing, 8*(3), 206–217.

Kwon, S. K., & Chidambaram, L. (2000). *A test of the technology acceptance model: The case of cellular telephone adoption*. In Proceedings of the 33rd Hawaii International Conference on Systems Sciences, Hawaii.

Legris, P., Ingham, J., & Collerette, P. (2003). Why do people use information technology? A critical review of the technology acceptance model. *Information and Management, 40*(3), 191–204.

Lu, J., Yu, C.-S., Liu, C., & Yao, J. E. (2003). Technology acceptance model for wireless Internet. *Internet Research: Electronic Networking Applications and Policy, 13*(3), 206–222.

McCreadie, C., & Tinker, A. (2005). The acceptability of assistive technology to older people. *Ageing & Society, 25*, 91–110.

Mokhtari, M., Khalil, I., Bauchet, J. Zhang, D., & Nugent, C. D. (2009). Ambient assistive health and wellness management in the heart of the city. In *Proceedings of the 7th International Conference on Smart Homes and Health Telematics, ICOST 2009* (Lecture Notes in Computer Science, Vol. 5597). Berlin: Springer.

Mynatt, E. D., Melenhorst, A.-S., Fisk, A. D., & Rogers, W. A. (2004). Aware technologies for ageing in place: Understanding user needs and attitudes. *Pervasive Computing, 3*(2), 36–41.

Nyseveen, H., Pedersen, P. E., & Thorbjørnsen, H. (2005). Intentions to use mobile services: Antecedents and cross-service comparisons. *Journal of the Academy of Marketing Science, 33*(3), 330–346.

Ortiz, M., Oyarzun, D., Yanguas, J., Buiza, C., González, M. & Etxeberria, I. (2007). *Elderly users in ambient intelligence: Does an avatar improve the interaction?* In Proceedings of 9th ERCIM Workshop 'User Interfaces For All' (pp. 99–114).

Osterberg, L., & Blaschke, M. D. (2005). Adherence to medication. *The New England Journal of Medicine, 353*(5), 487–497.

Øvretveit, J. (1992). *Health service quality*. Oxford: Blackwell Scientific.

Phang, C., Sutanto, J., Kankanhalli, A., Li, Y., Tan, B., & Tep, H. (2006). Senior citizens' acceptance of information systems: A study in the context of e-government services. *IEEE Transactions on Engineering Management, 53*(4), 555–569.

Röcker, C., Ziefle, M., & Holzinger, A. (2011, July). Social inclusion in AAL environments: Home automation and convenience services for elderly users. In *Proceedings of the International Conference on Artificial Intelligence (ICAI'11)* (Vol. 1, pp. 55–59), Las Vegas, NV, USA.

Savikko, N., Routasalo, P., Tilvis, R., Strandberg, T. E., & Pitkälä, K. H. (2005). Predictors and subjective causes of loneliness in an aged population. *Archives of gerontology and geriatrics, 41*, 223–233.

Shafer, R. (2000). *Housing America's seniors.* Cambridge, MA: Joint Center for Housing Studies, Harvard University.

Sixsmith, A., Mueller, S., Lull, F., Klein, M., Bierhoff, I., Delaney, S., et al. (2010). A user-driven approach to developing ambient assisted living systems for older people: The SOPRANO project. In *Intelligent technologies for bridging the grey digital divide* (pp. 30–45). Hershey, PA: IGI Global.

Sixsmith, A., & Sixsmith, J. (2008). Ageing in place in the United Kingdom. *Ageing International, 32*(3), 219–235.

Starner, T., Auxier, J., Ashbrook, D., & Gandy, M. (2000). *The gesture pendant: A self-illuminating, wearable, infrared computer vision system for home automation control and medical monitoring* (pp. 87–94). In Proceedings of the IEEE International Symposium on Wearable Computing (ISWC'00).Washington, DC: IEEE Computer Society.

Sugisawa, H., Liang, J., & Liu, X. (1994). Social networks, social support, and mortality among older people in Japan. *Journal of Gerontology, 49*, S3–S13.

vanden Broek, G., Cavallo, F., & Wehrmann, C. (2010). *Ambient assisted living roadmap. Developed by AALIANCE – the European ambient assisted living innovation platform.* Amsterdam: OIS.

Venkatesh, V., & Morris, M. G. (2000). Why don't men ever stop to ask for directions? gender, social influence, and their role in technology acceptance and usage behaviour. *MIS Quarterly, 24*(1), 115–139.

Venkatesh, V., Morris, M., Davis, G., & Davis, F. (2003). User acceptance of information technology: Toward a unified view. *MIS Quarterly, 27*(3), 425–478.

Veryzer, R. (1995). The place of product design and aesthetics in consumer research. In F. R. Kardes & M. Sujan (Eds.), *Advances in consumer research* (Vol. 22, pp. 641–645). Association for Consumer Research: Provo, UT.

Walker, D., & Beauchene, R. (1991). The relationship of loneliness, social isolation, and physical health to dietary adequacy of independently living elderly. *Journal of the American Dietetic Association, 91*, 300–304.

Wilkowska, W. & Ziefle, M. (2011). *Perception of privacy and security for acceptance of E-health technologies: Exploratory analysis for diverse user groups.* Full paper on the Workshop User-Centred-Design of Pervasive Health Applications (UCD-PH'11), held in conjunction with the 5th ICST/IEEE Conference on Pervasive Computing Technologies for Healthcare 2011.

Wixom, B., & Todd, P. (2005). A theoretical integration of user satisfaction and technology acceptance. *Information Systems Research, 16*(1), 85–101.

Wolters, M., Campbell, P., DePlacido, C., Lidell, A., & Owens, D. (2007). *Making speech synthesis more accessible to older people.* In 6th ISCA Workshops on Speech Synthesis (SSW-6), Bonn, Germany.

World Health Organization. (2003). *Adherence to long-term therapies. Evidence for action.* Geneva: World Health Organization.

World Health Organization. (2010). *Global recommendations on physical activity for health.* Geneva: World Health Organization.

Yi, M. Y., & Hwang, Y. (2003). Predicting the use of web-based information systems: Self-efficacy, enjoyment, learning goal orientation, and the technology acceptance model. *International Journal of Human-Computer Studies, 59*, 431–449.

Chapter 9
The Virtual Environment in Communication of Age-Friendly Design

Eunju Hwang, Andrew Park, Andrew Sixsmith, and Gloria Gutman

9.1 Introduction

There is a mismatch between the design of many communities and the needs of older persons. Many communities are designed for people who are at work during the day and at home during the night. Their main travel method is by car. Typically the connectivity and walkability between housing and services, such as doctors' offices and grocery stores, have not been a priority. However, with the increasing number of older persons aging-in-place, many communities are in the process of major transformation of their environments to be more supportive and age friendly. To date, over 100 local communities across Canada are engaged in Age-Friendly Communities (AFC) activities and five provinces have identified AFC as a priority (PHAC, 2010).

Obviously, well-designed communities facilitate various types of activities in all age groups. There is evidence that certain built environmental attributes (e.g., availability of facilities, infrastructure design, street patterns, destinations, traffic, green space, and population densities) are associated with sport, physical, and community

E. Hwang, Ph.D. (✉)
Apparel, Housing, and Resource Management, Virginia Tech,
240 Wallace Hall (0410), Blacksburg, VA 24061, USA
e-mail: hwange@vt.edu

A. Park, Ph.D.
Computing Science, Thompson Rivers University,
900 McGill Road, Kamloops, BC, Canada
e-mail: Apark@tru.ca

A. Sixsmith, Ph.D. • G. Gutman, Ph.D.
Gerontology Research Centre, Simon Fraser University, Harbour Centre,
#2800-515 West Hastings Street, Vancouver, BC, Canada V6B 5K3
e-mail: Sixsmith@sfu.ca; gutman@sfu.ca

A. Sixsmith and G. Gutman (eds.), *Technologies for Active Aging*,   155
International Perspectives on Aging, DOI 10.1007/978-1-4419-8348-0_9,
© Springer Science+Business Media New York 2013

activities (Frank, Engelke, & Schmidt, 2003; Frank, Schmid, Sallis, Chapman, & Saelens, 2005; King, 2008; King et al., 2005; Lee, Cubbin, & Winkleby, 2007; Mota, Lacerda, Santos, Ribeiro, & Carvalho, 2007). However, these studies to date have failed to provide strong evidence regarding the benefits of environmental interventions in terms of increased levels of participation in sport, physical, and community activities.

A key problem in planning research is that experimental studies require the implementation and evaluation of actual infrastructure change. A virtual environment (VE) provides a potential solution to the problem and a useful approach to study interventions designed to promote active aging and participation. VE refers to a simulated, computer-generated, interactive environment of either a real or a hypothetical location. In urban planning, VE has been used to share ideas and to communicate with the public at early planning stages. The technology encourages public participation in the planning process (Decker, 1994) and computer-simulated animation tends to foster more communication compared with the traditional pencil-and-paper approach (Levy, 1995). The action-oriented nature of perception in navigation and communication has been emphasized since the early study of VE (e.g., Gibson, 1986). Additionally, many pencil-and-paper assessment tools have been criticized for inadequate reliability and ecological validity in contexts detached from the real environment (e.g., Rizzo, 2001). The major obstacle to making real-world design tools has been the cost of these projects. Nevertheless, VE offers a useful tool for demonstrating design features that facilitate or impede active participation and identifying user preferences before actual implementation of an infrastructure change. Changes of perceptions, design ideas, and solutions can be reflected in the various developmental stages when VE is used.

A VE tool can also be used for assessing the built environment when navigating a real one is challenging or problematic. As we grow older, our environment has a direct impact on our quality of life (Lawton, 1986). A number of barriers to walking have been identified for older persons including the lack of sidewalks, poor maintenance of sidewalks, poor lighting, and increased street crime (CFLRI, 2004). These factors are particularly relevant to older persons living in low socioeconomic status (SES) neighborhoods (which are very often high crime areas) and in which physical activity levels are found to be significantly lower compared to those of medium to high SES neighborhoods (Parks, Housemann, & Brownson, 2003). In low SES neighborhoods, the major barriers, resulting in low physical activity levels, include poor-quality street maintenance and vandalism (Heinrich, Lee, Regan, et al., 2008; Mota et al., 2007) and, as previously mentioned, crime. Taking older people to such potentially fear-generating real environments to test potential design interventions can be problematic and unethical. Computer-generated VE enables them to take a virtual walk there. While navigating the VE, it is possible to observe the interaction between participants and selected elements of the physical environment that might be altered to creat a more age-friendly community.

There is a clear need to study whether changes to the built environment are likely to result in behavioral variance before expending large amounts of funding to implement them. Older adults' daily activity patterns need to be studied in the context of low SES neighborhoods and cultural settings. In addition, the main concept of AFC is that the perspectives of older adults are valuable at all planning stages. It is now required that their voices be heard and that their needs are reflected in the decision-making process and local planning (BC Ministry of Health Services, 2010). The next section of this chapter shows how a VE has been developed and used in a low SES, ethnically concentrated urban neighborhood to achieve these goals. Our specific objectives were to identify the key AFC aspects in the selected neighborhood, to evaluate the built environment of the neighborhood, and to develop a VE for communicating seniors' input at all the community planning stages.

9.2   Study Site

The site is in Vancouver's downtown Chinatown area. Protected as a Historic Area District, the Chinatown area is the oldest neighborhood in Vancouver. The core part of Chinatown is two kilometers extending from the intersection of Pender Street and Columbia Street and bound by Carrall, Gore, Union, and Hastings Street (City of Vancouver, 2009). Chinatown is located very close to the Downtown Eastside of Vancouver, an area associated with the drug trade, drug use, and high crime rates (Anderson, 2007). This is the poorest neighborhood in Vancouver and all of Canada (Walks & Maaranen, 2008). At the same time, this is one of the neighborhoods with the highest percentage of seniors in Metro Vancouver (City of Vancouver, 2010). The site also offers various environmental encounters such as different destinations and crossing roads and has gone through the Chinatown Community Plan (CCP):
(http://vancouver.ca/commsvcs/planning/chinatown).

One of the priorities included in the CCP has been to make Chinatown more age-friendly.

9.3   Procedure

To identify community needs and AFC priorities, we used mixed methods: in the first phase of the study, a focus group interview; in phase two, environmental audits; and in phase three, oral narratives.

9.3.1   Focus Group

The purpose of the focus group was to learn about any recent general and AFC activities in Chinatown. The participants, nine in total, were comprised of older residents of Chinatown, service providers, volunteers, and retired planners. The opening question was "What is an age-friendly community?" This was followed by questions that probed their perception of the most important aspects of AFC and explored what has been done to promote AFC in the Chinatown area. The focus group interview took two hours and was facilitated in Chinese and translated into English later.

Fig. 9.1 Study site, Chinatown, Vancouver, BC

9.3.2 Environmental Audits

From the focus group interview, we learned that core activities for seniors happen between the United Chinese Community Enrichment Services Society (S.U.C.C.E.S.S.) Community Center and the T&T Supermarket (see Fig. 9.1) where four seniors housing projects and local amenities are located. Using the Seniors Walking Environmental Audit Tool-Revised (SWEAT-R), we assessed walkability of seven street segments between the S.U.C.C.E.S.S. Community Center and the T&T Supermarket which are the two most frequently visited places among the Chinese senior residents of Chinatown. The SWEAT-R is a walkability assessment tool for use at the street level. Items on the SWEAT-R cover four topic areas (Michael, McGregor, Chaudhury, Day, Mahmood & Sarte, 2009):

1. Functionality (e.g., land use, building types, sidewalks, and buffer zones)
2. Safety (e.g., lighting and traffic safety)
3. Aesthetics (e.g., quality of microscale urban design, visual appeal of streetscape)
4. Destination (e.g., availability of senior-oriented housing and services/amenities, retail stores, transportation, and parking in the neighborhood)

Existence (1) or not (0) of each item was explored at each segment. Pictures of buildings and streets were also taken at each segment and we kept observation notes regarding detail materials, traffic volume, and social interaction at public places.

9.3.3 Oral Narratives

In the third phase, using VE, we obtained oral narratives from older adults living in Chinatown to identify their needs and experiences in their own voices. The oral narratives were used in conjunction with the environmental audits to reflect the older adults' perspective. A similar approach using both VE and street audits was used by Walford et al. (2011) and found to be very effective.

Constructing the VE

Using a 3D modeling software package (Maya), 3D models of the buildings, streets, trees, people, and all the detailed objects including pavements, curbs, signs, street furniture, and coin parking meters were built. Using a satellite picture, all the buildings and other objects were placed at their correct location. The pictures taken during the street audits provided the textures to construct the VE. After applying these onto the 3D models, lights were set up with proper intensity and locations. Then, the simulated 3D models between the S.U.C.C.E.S.S. Community Center and the T&T Supermarket were imported to a game engine (Unity 3D) so that participants could interact with the VE. A Nintendo Wii remote controller was used as an input device to experience the virtual walk.

Participant Recruitment

The participants for the oral narratives were recruited through the S.U.C.C.E.S.S. Community Center. A total of 21 older adults participated. They ranged in age from 65 to 88; 11 (52 %) were male and 10 (48 %) were female.

The project was advertised through S.U.C.C.E.S.S.'s community bulletins and notices were posted in public places in the Community Center indicating the time and location of the study. The staff at the S.U.C.C.E.S.S. Community Center made initial contact with potential participants, informed them about the study, and asked for their consent to be contacted by the researchers. All of the participants had no previous experience using VE.

Procedure

The oral narratives took place in a room at the S.U.C.C.E.S.S. office in Chinatown. They were facilitated in Chinese and translated later into English. Before each narrative, the participant was asked to rate his/her health, physical mobility, and level of performance of activities of daily living (ADL) and instrumental activities of daily living (IADL). The VE was played on a 51 in. LCD TV and the distance between the TV screen and participant was 2 meters. An individual session had the following three steps and lasted 30–90 minutes:

(1) Training: Participants had to learn how to use the Wii remote controller. The participants explored the text-run VE that was designed to display how to use the Nintendo Wii remote controller. Once they were sufficiently familiar with using the remote controller, an individual narrative using the VE started.

(2) Oral narratives: Each participant was asked to navigate the VE from the S.U.C.C.E.S.S. Community Center to the T&T Supermarket which was two blocks from S.U.C.C.E.S.S. While navigating, their route choices were observed and recorded. Participants' navigation was scored in terms of how well they proceeded to the destination and managed at decision-making points and junctions. Participants were asked to talk about the reasons for their choices and to describe features of the walking environment (e.g., buildings, streets, signs, bus stops) and how they felt along the way (e.g., likes and dislikes at different stages of the walk). The oral narratives were videotaped and later transcribed.

(3) Post-narrative interviews: After each narrative, there was a post-narrative interview during which we collected participants' demographic information and asked about their overall experience using the VE (Figs. 9.2 and 9.3).

Fig. 9.2 Virtual environment

Fig. 9.3 Participant navigating the virtual environment

9.4   Results

9.4.1   Focus Group

Several themes emerged from the focus group interview:

(1) Access to various destinations is important:

> In Chinatown, government provides low-rental housing or subsidized housing....There are also many societies, restaurants, S.U.C.C.E.S.S. and Chinese Cultural Centre. There are many resources here and it is very convenient.

> Age Friendly Communities should have good access to public transportation. It's more convenient to be close to bus stops.

(2) Safety has to be addressed:

> A lot of people do not want to live in apartments in Chinatown because the area is too threatening.

> If the area is not safe and things always happen, seniors would be afraid of it... if the area is not safe, of course no one would like to go there.

> People are afraid of going outside.

(3) People need to go outside:

> People need to go out and meet friends, do shopping and go to restaurants.

> (If people don't go outside), seniors are really depressed. They often cry. That's miserable.

> It's good to live somewhere easy to go to restaurants and stores especially for those who are disabled. They can't walk themselves. If…someone takes them outside and takes a walk in the mall, they would be healthier. It also reduces the medical expenses. Seniors don't feel happy at home all the time.

9.4.2 Street Audits

The two blocks between the S.U.C.C.E.S.S. Community Center and the T&T Supermarket included four walking routes: via Pender-Abbott Street which was very busy with high traffic volume and shops; via Carrall-Keefer Street where a lot of graffiti and vandalism was observed but parks were present; via Taylor-Keefer Street which was a relatively quiet shopping street; and through Shanghai Alley which was a very quiet narrow back alley (Figs. 9.4, 9.5, 9.6, and 9.7).

Fig. 9.4 Pender Street

Fig. 9.5 Carrall Street

Fig. 9.6 Taylor Street

Fig. 9.7 Shanghai Alley

Functionality

The street audits showed that except for the walking route via Shanghai Alley, there was mixed land use. There were sidewalks in all four walking routes although the width of sidewalks varied between 4 and 6 feet. With the exception of the route via Carrall Street, all the sidewalks were covered by awnings.

Aesthetics

The presence of benches and the degree of maintenance of buildings contribute to aesthetics. Benches enable older persons to take a rest during their walk and function as street decoration as well. The only usable benches observed were on the walking route via Carrall Street. On other segments, litter, graffiti-covered, and damaged benches were observed (e.g., Pender and Carrall).

Safety

Curb cuts were observed to be two-sided at all the walking routes, but not all the intersections were marked for pedestrian crossing. In addition, stop and yield signs were present only at the routes via Carrall-Keefer and Pender-Abbott.

Destination

Destination is associated with places for seniors, gathering places, retail shops, and public spaces. Except for the route via Pender-Abbott, other segments included seniors housing and seniors gathering/service places. The routes through Taylor-Keefer and Pender-Abbott presented many destinations (e.g., restaurants, retail shops, galleries). The Carrall-Keefer route included two parks and the Taylor-Keefer route presented two plazas. Bus stops were present at all the walking routes except Shanghai alley.

9.4.3   Oral Narratives

The oral narratives, captured while each participant walked virtually between the S.U.C.C.E.S.S. Community Center and the T&T Supermarket, were coded based on the SWEAT-R categories from the street audits. Keywords were highlighted at each decision-making point (i.e., intersection). Walford, Samarasundera, Phillips, Hockey, and Foreman (2011) used this approach to relate more closely to the everyday language used by the public. We adapted their approach because it would reflect older persons' own voices and experiences.

While the participants walked virtually using the VE, we asked them to show us how to get to the T&T Supermarket from the S.U.C.C.E.S.S. Community Center. Of the successful attempts to reach the final destination ($n = 19$), 9 participants selected the route through Taylor-Keefer and 8 participants selected the route through Pender-Abbott. The participants talked freely about what they were exploring, what motivated them to make their choice of route, and about their likes and dislikes.

Safety was a major concern for many participants. While it is affected by built environmental features (e.g., quality of sidewalk, signage, curb cuts, marked crossings, traffic volume), many participants mentioned that they felt safer when there were more people and more cars present:

> I will choose Abbott because it usually has more people and cars which make me feel safer.

> [I] use Taylor street because it's busier and the shorter to T&T. [Shanghai] Alley is too quiet to walk through which makes me feel unsafe.

> I don't use Shanghai Alley because I don't like small alley. There (are) no people around and the street is not clean.

There was no commentary on signage, the need to make 90 degree turns, or about curb cuts, but threatening behavior including the presence of homeless people was mentioned:

> Carrall street has homeless people hang around. [I] don't want to take that road.

> I never walk in [Shanghai] alley because I have seen beggars and drug users in alley.

Destination

Many participants extensively explored Pender-Abbot and Taylor-Keefer in order to reach T&T. They preferred the routes with more shops and local amenities, which created a more interesting walking environment:

> [I] usually walk through the International Village[1] to reach T&T because I can shop on the way to T&T.

> Usually I walk through the International Village because it's safer and I can shop around or go get some food/coffee...

> If the mall is open, I will walk through the mall because I can shop around. If the mall is not open, I will use Abbott because there used to be art exhibition.

> Usually I walk through the International Village to T&T because there is more stuff to look at....

Functionality

Several participants mentioned the width and quality of sidewalks:

> I chose Taylor-Keefer because there is sidewalk and the space is bigger. Usually I use Taylor because it's clean.

> [I] use Taylor because I wanted to use the pavement... roads are wider.

9.5   Discussion

Engagement of older persons in AFC planning is essential. The goal of AFC is to empower older people to live in security and good health and continue to participate fully in society. For this, they need to be an integral part of community planning and any decision-making process. It is true that a conventional map can provide useful information in AFC studies. However, it has been reported that much more effective and intuitive understanding occurs when images and data are provided in 3D (Daniel, 1992). Nevertheless, the role of VE as a tool for AFC research has not been widely addressed. The study described in this chapter has been conducted parallel to development of the Chinatown Community Plan, in partnershipwith S.U.C.C.E.S.S.

The VE was explored as a new method of engaging older persons in Chinatown in identifying their own needs and priorities to create a more age-friendly Chinatown. Due to the significant interests in the area (in relation to the city's Eastside Economic Revitalization project and Vancouver's hosting of the 2010 Winter Olympic games), considerable amounts of spatial data had been compiled in a digital form, and this provided material to construct a VE of sufficient detail in a relatively short period of time.

---

[1]A three-story shopping mall identified as Tinseltown in Fig. 9.1 that includes a movie theater and food court on the top floor, pharmacy, McDonalds, Starbucks, 7–11, and other fast-food outlets on the main floor as well as retail shops selling clothing and other goods and services.

Navigating through the VE is important because it represents the real environment that we live in. Visualizations, however, also produced deviated and subjective impressions in the study participants. While the VE aimed to enable people to evaluate their familiar environment, some of the participants were unable to convey information on the walking route that they used frequently. Subtle height differences, depth, and distance estimates were misinterpreted by those who could not reach the destination in less than two attempts. In addition, subtle height differences and slopes were not easy to identify. Also, the safety categories of the SWEAT-R address features of the built environment relevant to mobility impaired persons. In the present study, the participants' narratives primarily reflected concern about the potential threatening behavior of homeless people and drug users. Interestingly, none of the cultural heritage features and landmarks (e.g., Shanghai Alley, Millennium Gates, Sun Yat-Sen Chinese Garden) were identified by the participants as a key factor in decision-making on walking routes. However, some did speak about quality of sidewalks and traffic volume.

Simulator sickness was assessed through self-ratings of dizziness. Participants did report some degree of dizziness when using the VE. The strong forward moving behavior using the VE may have made it difficult for participants to explore their options at the intersections (decision-making points). This might lead to making decisions on their walking route differently compared to the traditional approach of using still maps. However, no urgent symptoms were reported during the oral narratives.

In conclusion, the study presented in this chapter shows the potential of computer-generated visualizations as a resource for age-friendly design communication. Since computer graphics are becoming more advanced, there is more realism to them. Therefore, before environmental changes are implemented in the community, VE should be considered as a way of determining people's perceptions and their opinions of it. The great value of VE is in its potential for participatory research, allowing local people to be involved in the planning process at an early stage.

With respect to Vancouver, the study site was part of the Chinatown Economic Revitalization Action Plan. The city has proposed to create more seniors housing, to renovate buildings, to improve the physical appearance of buildings, and to design streets that promote walkable and safe communities: (http://vancouver.ca/commsvcs/planning/chinatown/pdfs/11nov28aecomreport. pdf). The study demonstrates a potential tool for gathering input from residents that hopefully will be used as the project moves forward.

References

Anderson, M. A. (2007). Location quotients, ambient populations and the spatial analysis of crime in Vancouver, Canada. *Environment and Planning, 39*, 2423–2444.

British Columbia Ministry of Health Services (BCMHS). (2010). *Age-friendly British Columbia*. Vancouver, Canada: Author.

Canadian Fitness and Lifestyle Research Institute (CFLRI). (2004). *A municipal perspective on opportunities for physical activity*. Retrieved May 20, 2009 from http://www.cflri.ca/eng/statistics/surveys/documents/2004capacity.pdf

City of Vancouver. (2009). *Chinatown historic district.* Retrieved July 15, 2011 from http://vancouver.ca/commsvcs/planning/chinatown/pdfs/ChinatownNHS_Nomination.pdf

City of Vancouver. (2010). *Seniors in Vancouver.* Retrieved May 19, 2012 from http://vancouver.ca/commsvcs/socialplanning/initiatives/seniors/pdf/Seniors_Backgrounder.pdf

Daniel, T. C. (1992). Data visualization for decision support in environmental management. *Landscape and Urban Planning, 21,* 261–263.

Decker, J. (1994). The validation of computer-simulations for design guideline dispute resolution. *Environment and Behavior, 26,* 421–443.

Frank, L. D., Engelke, P. O., & Schmidt, T. L. (2003). *Health and community design: The impact of the built environment on physical activity.* Washington, DC: Island.

Frank, L. D., Schmid, T. L., Sallis, J. F., Chapman, J., & Saelens, B. E. (2005). Linking objectively measured physical activity with objectively measured urban form. *American Journal of Preventive Medicine, 28,* 117–125.

Gibson, J. J. (1986). *The ecological approach to visual perception.* Hillsdale, NJ: Lawrence Erlbaum Associates.

Heinrich, K. M., Lee, R. E., Regan, G. R., et al. (2008). How does the built environment relate to BMI and obesity prevalence among public housing residents? *American Journal of Health Promotion, 22,* 187–194.

King, D. (2008). Neighborhood and individual factors in activity in older adults: Results from the neighborhood and senior health study. *Journal of Aging ad Physical Activity, 16*(2), 144–170.

King, W. C., Belle, S. H., Brach, J. S., Simkin-Silverman, L. R., Soska, T., & Kriska, A. M. (2005). Objective measures of neighborhood environment and physical activity in older women. *American Journal of Preventive Medicine, 28*(5), 461–9.

Lawton, M. P. (1986). *Environment and aging.* Monterey, CA: Brooks/Cole.

Lee, R. E., Cubbin, C., & Winkleby, M. (2007). Contributions of neighborhood SES and physical activity resources to physical activity in women. *Journal of Epidemiology and Community Health, 61,* 882–890.

Levy, R. M. (1995). Visualisation of urban alternatives. *Environment and Planning B: Planning and Design, 22,* 343–358.

Michael, Y. L., McGregor, E. M., Chaudhury, H., Day, K., Mahmood, A., & Sarte, A. (2009). Revising the senior walking environmental assessment tool. *Preventive Medicine, 48,* 247–249.

Mota, J., Lacerda, A., Santos, M. P., Ribeiro, J. C., & Carvalho, J. (2007). Perceived neighborhood environments and physical activity in an elderly sample. *Perceptual & Motor Skills, 104,* 438–444.

Parks, S. E., Housemann, R. A., & Brownson, R. C. (2003). Differential correlates of physical activity in urban and rural adults of various socio-economic backgrounds in the United States. *Journal of Epidemiology and Community Health, 57,* 29–35.

Public Health Agency of Canada (PHAC) & Agence de la sante publique du Canada (ASPC). (2010). *Healthy living eBulletin-theme: Age-friendly communities.* Retrieved June 10, 2010 from http://www.phac-aspc.gc.ca/hl-vs-strat/e-bulletin-eng.php

Rizzo, A. (2001, March 13). *Advances in the application of virtual environments for mental healthcare.* In IEEE, Yokohama, Japan.

Walford, N., Samarasundera, E., Phillips, J., Hockey, A., & Foreman, N. (2011). Older people's navigation of urban areas as pedestrians: Measuring quality of the built environment using oral narratives and virtual routes. *Landscape and Urban Planning, 100,* 163–168.

Walks, A., & Maaranen, R. (2008). *Neighbourhood gentrification and upgrading in Montreal, Toronto and Vancouver.* Retrived April 29, 2012 from http://www.urbancentre.utoronto.ca/pdfs/researchbulletins/CUCS_RB_43-Walks-Gentrification2008.pdf

Chapter 10
Technology and Fun for a Happy Old Age

Arlene Astell

10.1 Introduction

The past 20 years have seen a rise in the development and production of technologies to support older people. These have typically focused on issues related to safety and security and to reduce the risk of hospitalization (e.g., fall detection devices). Despite their undoubted importance, it could be argued that these aspects of aging have received more attention than is warranted and as a consequence have unduly influenced the direction of technology development for the aging population. While much less attention has been paid to technology to support people to live well and experience the things that make life worth living, the evidence that is available suggests that technology can provide people with meaningful and engaging activities that are stimulating, enjoyable, and fun. This chapter provides a brief examination of this evidence for the aging population in general and then considers the application of technology for that sector of the aging population who are living with dementia. The context is provided by positive psychology, an approach to human behavior that seeks to promote the good things in life.

10.2 Positive Psychology

While most people, whatever their age, are concerned with staying safe and well, there is more to life than safety and security. In his oft-cited *Hierarchy of Needs*, Maslow (1943) identified meeting our safety needs as an important but very basic human need, second only to satisfying our physical needs such as hunger and thirst.

A. Astell (✉)
CATCH (Centre for Assistive Technology and Connected Healthcare),
School of Health and Related Research, University of Sheffield, Sheffield S1 4DA, UK
e-mail: a.astell@sheffield.ac.uk

A. Sixsmith and G. Gutman (eds.), *Technologies for Active Aging*,
International Perspectives on Aging, DOI 10.1007/978-1-4419-8348-0_10,
© Springer Science+Business Media New York 2013

However, central to Maslow's theory of human motivation was the belief that humans seek to fulfill higher level, more complex psychological needs, such as the need for satisfying relationships, achievement, and the respect of others. His *positive theory* was predicated on seeing the person as a whole within the environment in which they operate and was primarily concerned with identifying and exploring those drivers of human behavior that go beyond meeting our basic survival needs.

While the resonance and endurance of Maslow's ideas are easy to find in writings and innovation across a wide range of fields in human creativity over the past 70 years, it is more difficult to see his influence in attitudes towards aging and, specifically, in meeting people's needs for a happy and successful later life. This is particularly so in the arena of technology development, which, as indicated above, has tended to emphasize the safety of older people, with developments such as alarm pendants that people wear and should press if they have a fall or the installation of unobtrusive or passive sensors in people's homes to alert external agencies of unusual events and trigger assistance if required. While these can be seen as good-intentioned moves, they focus on only the most basic of human needs, suggesting either (1) a belief that older people do not have the same higher level psychological needs as younger people or (2) that older people do have the same higher level needs but that meeting them is either (a) less important or (b) more difficult than it is for younger people and therefore meeting the basic survival needs of older people should suffice.

The apparent ignorance or dismissal of older people's more complex needs is surprising given the efforts over the past 20 years of the positive psychology movement, which grew from and to a large extent embodies not only Maslow's work but also that of other influential writers on the broad subject of human happiness. Positive psychology is concerned with improving people's lives by nurturing them to thrive and flourish. Positive psychology developed in part as a response to a growing concern that psychology as a discipline was becoming increasingly negative, focusing too much on mental ill health and disorders of human existence (Seligman & Csikszentmihalyi, 2000). While these are a major and growing problem in society, mental disorders do not account for the majority of people's experience for the majority of their lifetime. Indeed, focusing instead on the conditions that promote *good* mental health and well-being should not only encourage people to live happier and more fulfilling lives but also help to prevent the occurrence of mental ill health, which is the goal of positive psychology.

10.3   Positive Psychology and Aging

In 1948 the Preamble to the Constitution of the World Health Organization defined health as "…a state of complete physical, mental and social well-being and not merely the absence of disease or infirmity" (p. 100), a definition that has remained unchanged. Positive psychology is concerned with the aspects and qualities that promote and provide a positive and meaningful existence—essentially the things that make life worth living. Concordant with Maslow's (1943) view, positive psychology considers the human experience from the individual to the social group, highlighting the "valued subjective experiences [of] well-being, contentment and satisfaction (in the past); hope and optimism (for the future); and Flow and happiness (in the present)" (Seligman & Csikszentmihalyi, 2000, p. 1). This approach recognizes that people do not exist and operate in isolation and thus the satisfaction of an individual's needs occurs within the social context, acknowledging the importance of interaction and relationships with others in achieving this.

There is no reason to believe that satisfying our human needs for well-being, achievement, hope, etc. lessens as we age. Additionally, there is a strong preventive argument to be made for assisting and supporting people to keep experiencing a meaningful and fulfilling life as they age for the benefits this can bring. In a study of successful aging Vaillant and Mukamal (2001) located older men on a continuum from *happy-well* to *sad-sick* based on their scores in six domains (1) objectively measured physical health (including absence of irreversible physical disability), (2) subjective physical health (completing activities of daily life), (3) length of active life, (4) objective mental health, (5) subjective life satisfaction, and (6) social supports (including relationship satisfaction). The *happy-well*, who scored highly in the six domains of successful aging, lived longer and had fewer years of disability than their counterparts who were judged to be both physically unhealthy and emotionally negative.

The need for interventions that promote positive aging and tackle negative aspects of older people's lives is further supported by the impact of depression. Although the most common mental illness, depression in older people is consistently under-diagnosed and under-treated (Age Concern, 2008), despite its links to increased disability and physical illness (Anderson, 2001). In the UK up to one in four community-dwelling older people—approximately two million people—has symptoms of depression that warrant treatment, with the risk increasing with age from one in five in the 65–69 age group to two in five in the 85+ age group (Anderson, 2001). As well as increasing vulnerability to other conditions in old age, depression can also interfere with the successful treatment of physical illness, as low mood and poor motivation make it difficult for people to engage with any kind of therapy (Salzman, 1995).

There is also a strong link between depression and suicide with figures from the United States, suggesting that older adults are disproportionately likely to die by suicide. In 2007, 14.3 per 100,000 people aged 65 and older died by suicide compared to the national average of 11.3 suicides per 100,000 people in the general population (National Institute of Mental Health, 2007). The figure was significantly higher for non-Hispanic white men age 85 or older at 47 suicide deaths per 100,000. Thus, finding ways and means to promote and support positive aging should be seen as a priority, not a luxury to be attempted only after physical and safety needs have been dealt with.

Technology can potentially assist people, young or old, to live a happy life. Indeed there are whole areas of the leisure technology industry dedicated to this, but the emphasis has tended to be on the younger generations, especially with developments in gaming and equipment for making, playing, and enjoying music. However, this is arguably the wrong sector to focus on as increasingly the consumers of technology will be the over-50s who have higher disposable income which they are looking to spend either on themselves or their aging relatives (Coughlin, 2010).

With this in mind, what follows is a brief and necessarily selective consideration of technology that has and is being developed and used to support older people to enjoy life. The examples are confined primarily to computer technology in the areas of games and social interaction/communication to give a flavor of what can be done with technology to meet the needs of older people. The information is divided into two sections: the first concerned with the aging population generally and the second relating to meeting the needs of older people with a diagnosis of dementia. Recognizing and responding to the increasing numbers of older people with cognitive impairment presents an additional but growing challenge that must be taken into account not only by designers and developers of future technologies but also by those planning and delivering services for the aging population.

10.4 Flow Experience and Games

What makes gaming enjoyable? Why do some people enjoy computer games and other people enjoy chess, cycling, rock climbing, learning a foreign language, or playing a musical instrument? Essentially these are activities that encourage mastery and achievement and provide an opportunity for people to develop and improve a skill as well as achieving satisfaction and a sense of accomplishment. This fits well with the positive psychology notion of *Flow*, whereby one experiences immersion in an activity for its own sake with the result that one feels a sense of satisfaction and loses track of time (Csikszentmihalyi, 1990).

Csikszentmihalyi (1990) defined a Flow experience as having eight dimensions (Table 10.1), which make it reinforcing and fulfilling and which create a desire for a person to repeat the experience. Among these are clear goals and immediate feedback plus a sense of potential control. For an activity to be self-rewarding and to "…maintain a person's Flow experience, the activity needs to reach a balance between the challenges of the activity, and the abilities of the participant…If the challenge is higher than the ability, the activity becomes overwhelming and generates anxiety. If the challenge is lower than the ability, it provokes boredom" (p. 111).

Essentially, Flow can be seen as defining the conditions for positive, meaningful experience, and playing computer games provides an example of how this can be achieved using technology. This has been embodied in the Presence-Involvement-Flow Framework (Takatalo, Nyman, & Laaksonen, 2008) of game playing, which attempts to characterize the particular elements of games that make them enjoyable and that encourage people to keep playing, including the important balance between ability and challenge.

Table 10.1 Csikszentmihalyi's (1990) eight dimensions of Flow

Dimension
1. We confront tasks we have a chance of completing
2. We must be able to concentrate on what we are doing
3. The task has clear goals
4. The task provides immediate feedback
5. One acts with deep but effortless involvement that removes from awareness the worries and frustrations of everyday life
6. One exercises a sense of control over their actions
7. Concern for the self disappears, yet, paradoxically, the sense of self emerges stronger after the flow experience is over
8. The sense of duration of time is altered

10.5 Aging and Computer Games

If games are fun and playing them can make us happy, are there particular types of games or other considerations that make them more or less appealing to older people? In an early review of the literature, Whitcomb (1990) found that although only a limited number of games had been investigated at that time, it was possible to identify a range of benefits for older people from playing computer games. These included the recreational pleasures of satisfaction and accomplishment which positively influenced people's view of themselves and their abilities. Whitcomb also noted physical and cognitive benefits including enhanced motor skills, such as hand-eye coordination and manual dexterity, increased speed on the games played, plus anecdotal evidence of these skills transferring to other aspects of people's daily lives, such as driving.

Building on these findings, Ijsselsteijn, Nap, de Kort and Poels (2007) identified four potential areas for games to contribute to improving the quality of life for older people. These they characterized as (1) relaxation and entertainment, (2) socializing, (3) sharpening the mind, and (4) more natural ways of interacting. Of these, relaxation can be seen as the enjoyable, rewarding aspects of gaming that are shared by people of all ages. The second and third elements, those concerned with social interaction and cognitive stimulation, are examined further below. The final element proposed by Ijsselsteijn et al. (2007) refers to developments such as the Sony EyeToy and Nintendo Wii™ that allow for more physical ways to engage with computer games, which may encourage people to try physical activity and fitness programs.

10.6 Games and Physical Activity

A current example is Wii Fit, which comprises more than 40 activities such as yoga, strength building, aerobics, and balance games. The system has proven remarkably popular with consumers of all ages although evidence about any potential health benefits is limited. For example, the aerobic activities were judged insufficient for maintaining target heart rate required for cardiorespiratory fitness in a study of different age groups although participants enjoyed the Wii activities more than walking or jogging on a treadmill (Graves et al., 2010). These findings suggest that presenting interactive exercise and health-related activities on a console has potential for encouraging people to engage with novel activities, even if currently available activities do not confer great health benefits.

This is supported by a case study examining the potential of the Wii Fit balance platform to contribute to rehabilitation of an older person with a history of falls (Pigford & Andrews, 2010). In this study the performance of an 87-year-old lady on balance measures improved through a traditional balance-training program combined with selective and monitored usage of the balance platform. The authors commented on specific benefits of the interactive environment, including presentation of goals and level of challenge, which motivated the player to try again, important factors in the Flow experience.

10.7   Games and Social Interaction

Alongside a potential contribution to physical health, many computer games have an intrinsic social element, being played with one or more other people. In his 1990 review, Whitcomb identified social interaction as the primary benefit that older people gain from playing games. More recent projects, such as Eldergames (Gamberini et al., 2007), have attempted to build on these early findings about the potential for games to benefit older people, coupled with increasing knowledge about neural plasticity and the capacity of the brain to compensate for age-related loss, to use games as a medium for more generally improving older people's quality of life. Using a model of user acceptance based on the theory of Flow, Eldergames examined participants' experience of their novel games table along seven key dimensions including social interaction, playability/immersion, challenge/skills, and clear goals. Participants identified social interaction, defined as the opportunity to create and maintain new relationships, as the biggest benefit in Eldergames, with 66 % endorsing the statement "The most interesting thing has been to share my time with other people while playing" (Gamberini et al., 2009, 167). The social benefits from playing computer games can also be delivered through playing remotely, such as in the Age Invaders' System, which was developed to encourage intergenerational activities online (Khoo, Merritt, & Cheok, 2009). Age Invaders was designed specifically to facilitate intergenerational family entertainment by enabling one or two different generations, for example, grandparents and grandchildren, to play games together while in different locations. This highlights not only the potential for exploiting the Internet for increased engagement and social interaction for older people but also the willingness and interest among the aging population to engage with technology. Older people's interests in technology, and their enthusiasm for adopting it, are significantly influenced by its potential for enhancing or increasing opportunities for social contacts, putting them in touch with people with similar interests or enabling them to stay in touch with people when face-to-face interactions are not possible, perhaps through distance or illness (Ijsselsteijn et al., 2007).

10.8 Aging and Cognitive Exercise

Alongside the direct personal benefits of engaging in an enjoyable and fulfilling activity, having fun and social interaction, game playing is also becoming increasingly popular for the possible benefits it confers through cognitive stimulation. There is increasing interest in the possibility that game playing can play a preventive role in resisting cognitive decline or even the onset of dementia in the aging population. This has arisen in part as a result of growing evidence that higher levels of cognitive activity may mitigate the impact of neuropathology in the brain. Most of this evidence has come from exploration of the apparent beneficial effects of higher levels/greater years of education on people whose brains are found to contain neuropathological changes at death but in whom there were minimal signs of dementia in life (Brayne et al., 2010). This is explained most readily by the notion of brain *reserve* (Stern, Alexander, Prohovnik, & Mayeux, 1992), whereby although education does not protect against the development of neuropathological changes per se, it does enable people to compensate or overcome the impact of neuronal loss through having greater cognitive reserves to draw on.

Explanations for why education results in this buffer against neuropathology have tended to relate to either biological processes such as synaptic density, whereby greater density will hypothetically be more resistant to the synaptic loss that occurs, for example, in Alzheimer's disease (Katzman, 1993), or functional processes relating to neural networks and improved compensatory activities by those with greater reserves (Stern, 2006). This raises the increasingly important question of whether other activities besides education can influence an individual's cognitive reserve and whether these can be undertaken later in life, with a view to offering some protection against decline in older age.

In a study of 469 older people recruited to the Bronx Study of Aging, Vorghese et al. (2003) found that those who developed dementia over a 21-year period not only had lower education levels, but they also scored significantly lower on scores of current cognitive activity. In this study cognitive activity comprised six activities (reading books or newspapers, writing for pleasure, doing crossword puzzles, playing board games or cards, participating in organized group discussions, and playing musical instruments), which participants carried out with varied frequency from daily to never. One striking finding was that participants "who did crossword puzzles four days a week (four activity-days) had a risk of dementia that was 47 % lower than that among subjects who did puzzles once a week (one activity-day)" (Vorghese et al., 2003, p. 2515). These findings lead the authors to suggest that advising older people to participate in cognitive activities could become as commonplace as recommending them to undertake physical activities to reduce the risk of cardiovascular disease.

Optimism that people can increase their own reserve through cognitive exercises or other stimulating cognitive activities comes from the life course approach to cognitive reserve proposed by Richards and Deary (2005). They argue that while some abilities such as problem-solving and logical thinking (so-called *fluid intelligence*) are particularly vulnerable to neuropathology, our abilities to use skills, knowledge, and experience (*crystallized intelligence*) are not only more resistant to neuropathology but can also be extended in adulthood. Evidence from young people who have grown up playing computer games suggests that the skills they have developed, including shutting out distractions and rapidly switching attention, may equip them to better resist the onset of disorders of aging (Bialystok, 2006). Perhaps engaging in similar gaming activities could confer similar benefits to older people.

10.9   Games and Cognitive Stimulation

In their 2009 review of digital action games, Zelinksi and Reyes identified *far transfer* as a key product of cognitive training for older adults. Pointing to the potential for extended practice training, they linked this to the elements, such as presence, engagement, and other elements of Flow, that make computer game playing a positive and immersive experience. Of particular interest is their dissection of the potential benefits of different types of digital action games for different types of cognitive function (Table 10.2), although they conclude that further research is required to quantify and determine the time frame for delivering these.

One recent development that seeks to take account of and recognize the different needs of young and older players is Age Invaders. As previously described this system, designed for intergenerational game playing, is interesting in that it takes into account both differences in game playing ability due to age and experience and changes in cognitive ability associated with age (Khoo et al., 2009). For example, in

Table 10.2 Hypotheses about abilities improved by different game genres

| Ability | Game genre | | | | |
	1st-person shooter	3rd-person action-adventure	Strategy	Role playing	Massive multiplayer
Eye-hand coordination	X	X	X	X	X
Memory			X	X	X
Mental rotation	X	X	X		
Reasoning			X	X	
Response speed	X	X			
Supervisory			X	X	X
Visual attention	X	Xa			
Working memory			X	X	X

x=the genre is hypothesized to produce significant improvements in performance; a=improvement expected in platformer games of this genre (Zelinksi & Reyes, 2009, p. 227) the Space Invaders game, the system allows older players more time to react to rockets fired by younger players. This accommodation may be important for facilitating the continuation of game playing as uneven pairings of a highly skilled and practiced player with a novice may discourage the learner, especially if they lack confidence about using technology.

Game playing also offers the potential for studying cognitive function and assessing changes in players' abilities—both gains and declines. Jimison, Pavel, Bissell, and McKanna (2007) have created a suite of computer games that includes metrics to facilitate detection of change within an individual over time, providing a means of detecting early cognitive decline as well as enabling the system to adapt the user interface to the needs of the individual. While still work in progress, this approach has potential wide-scale application both in terms of detection of cognitive decline and in developing systems that can respond appropriately to user needs, a critical feature for developing novel technologies that can support people to live as well as possible as their needs change.

Thus, playing computer games, whether traditional games such as Solitaire, Scrabble, or chess or newer games developed for dedicated gaming systems, can be seen to provide potential health, social, and cognitive benefits as well as being a rewarding and fulfilling activity for people of any age. Playing games also offers some additional benefits of particular importance to the aging population in being portable, an indoors pastime that mostly does not require a large amount of physical input, thus making it a suitable activity for people with reduced mobility and limited opportunities for activities outside the home, all important when considering the potential contribution of technology to making the lives of older people worth living.

10.10   Other Potential Benefits of Technology

The potential of computer technology to help motivate and support older people to positively manage their lives and make health-related behavior change has been explored in a number of other projects. For example, Jimison and Pavel (2007) reported a new model for delivering computer-based health interventions for older people in their own homes. Their system uses key principles of health behavior change including goals, motivation, and readiness to change, addressed through specific prompts and weekly progress checks with a professional health coach via a computer. The package is tailored to each individual to meet their particular needs and help them set and meet their own goals, with regular encouragement and progress checks from the health coach. The aim is to provide enjoyable activities and the right balance of motivation, support, and encouragement to enable people to exert control over their lives. The same team is currently exploring the potential of using a more immersive interactive video environment for coaching older people in exercises (Pavel, personal communication).

10.11   Technology and Dementia

A growing number of people in the aging population have a diagnosis of dementia. Dementia is an umbrella term applied to the occurrence of progressive and irreversible cognitive decline that causes difficulties carrying out daily activities such as shopping and banking, problems recalling people's names, or difficulties planning and completing journeys. With the passage of time, the difficulties people face with dementia increase to the point where they need significant help and support to manage their daily lives, highlighting a potential role for technology.

Given that the biggest risk factor for developing a dementia is advancing age (Alzheimer's Association, 2010), technology will become increasingly important as the falling birth rate results in fewer caregivers to meet the needs of the increasing numbers of older people we can expect to be living with a dementia diagnosis. As life expectancy increases, the number of people with dementia in the world is predicted to rise from 25 million in 2000 to 63 million by 2030 and to 114 million by 2050 (Alzheimer's Association, 2010). However, dementia is a difficult disorder to deal with, as there is no single cause and no single pattern to the way it affects people. Alzheimer's disease (AD), which accounts for approximately 42 % of cases, is the most common cause although many other people have vascular dementia (23.7 %) and mixed AD and vascular dementia (21.6 %) (Brunnström, Gustafson, Passant, & Englund, 2009). Latest estimates put the current cost of providing care for people with dementia in excess of $601 billion, accounting for 1 % of the world's Gross Domestic Product with an estimated rise of 85 % by 2030 (Wimo & Prince, 2010). This demands an urgent response as the numbers affected by dementia are predicted to keep rising. While the potential of technology to support people with dementia has recently been recognized, as with developments for the rest of the aging population, the bulk of the work to date has fallen almost exclusively into the safety and security category (Astell, 2006). These include the use of automated medication prompts and electronic tagging (Bail, 2003) for people with memory difficulties, some of which raise ethical issues in relation to their application with people with dementia. Even so, it could be argued that there is even greater potential for technology to assist people with dementia to live well than for the rest of the aging population if it is approached with a positive view of enabling people to live a fulfilled and enjoyable life.

10.12   Cognitive Training and Dementia

Outside of safety and security developments, to date only a small number of studies have investigated other applications of technology for people living with dementia. Of these a number have looked at the potential of computers to deliver training either to slow cognitive decline in people with dementia (e.g., Mate-Kole et al., 2007) or enable them to keep carrying out daily activities (e.g., Hofman et al., 2003).

In their study of five females and one male with moderate dementia, aged between 64 and 93 years old, Mate-Kole et al. (2007) examined the impact of a 6-week intervention comprising Mind Aerobics (an interactive group training seminar) combined with Adaptive Computerized Cognitive Training (ACCT), a program of activities targeting a range of cognitive skills including attention, memory and problem-solving that can be modified to the individuals' needs. In this preliminary study, significant improvement is reported in global cognitive function in five of the six participants plus gains in selective remembering for all participants at the end of training. The participants' scores on measures of daily activities also improved, supported by caregiver reports of changes in behaviour including increased awareness of the environment, increased socialization and initiation of interactions. Although small scale, this study suggests that computerized activities have something to offer older people who have a dementia diagnosis, not just in terms of cognitive function but in terms of practical transfer into the rest of their lives.

10.13   Computers and Daily Activities

A number of studies by Hofman and colleagues (Hofman, Hock, Kuller, & Müller-Spahn, 1996; Hofman, Hock, & Müller-Spahn, 1996; Hofman et al., 2003) have explicitly examined the potential of computerized training to support people with dementia to continue to carry out everyday tasks of functional relevance such as learning a route (e.g., finding the way to the bakery or to a caregiver's apartment in the same neighbourhood) or shopping for items on a list. In these small studies of four, nine, and ten participants, respectively, the tasks were individualized to the participants, for instance, by using photographs of their real apartment and neighborhood. In all three studies, participants' performance on the trained tasks improved, with evidence of training effects being retained 3 weeks later. This highlights not just the potential but also the importance of supporting people with dementia to continue to carry out daily activities, first by reducing the burden on family and formal caregivers to take over these tasks and second to maintain the skills of people living with dementia, which would be further undermined and potentially lost quicker than needs be, if caregivers carry out activities for them (Mate-Kole et al., 2007).

   Continued independence and autonomy is as important for people with dementia as it is to anyone else (Astell & Orpwood, 2010). Just as computer gaming provides the opportunity for both young and older players to enjoy a *Flow* experience, people with dementia could also benefit from the opportunity to engage in immersive and engaging activities. This requires that games and activities be designed and presented in a way that takes account of the progressive cognitive needs of people with dementia. While interactive systems such as the Nintendo Wii™ have attracted popular attention for their possible entertainment and even physical activity value for older people with dementia, such off-the-shelf systems and games are not designed to take account of the particular needs of people with dementia. Specifically, most commercially available products assume intact cognitive function to (1) permit players to understand the relationship between a handheld controller that remotely influences activity on a screen and (2) accomplish the learning and improvement on the activities that provide the motivation to play again. Given the prominent memory difficulties of people with Alzheimer-type dementia, it is difficult for them to explicitly learn new skills and tasks by the traditional means on which these games rely.

10.14   Engaging Activities for People with Dementia

The Living in the Moment project (Astell et al., 2009) was established to explore the parameters for developing computerized activities to provide engaging and stimulating activities for people with dementia to enjoy. Working solely with people with dementia, as opposed to formal or family caregivers, the project set out to identify what sorts of activities were most enjoyable for older people with dementia and how they could best be prompted to initiate and continue to play games independently. Utilizing touch screen technology, this involved the creation and testing of more than 30 activities from a 3D tour round a botanic garden (Fig. 10.1; Astell et al., 2009) to painting a pot (Fig. 10.2) to fairground games such as duck shooting (Astell, 2010).

Over a number of years the findings revealed that people with dementia can actually learn to interact with and play computerized activities through implicit means. Participants were able to use the system and interact with the computer independently in the absence of a caregiver. There was also evidence of learning as people improved in accuracy or speed, depending on the demands of the game (Astell, 2010). The findings suggest that there is great potential for developing stimulating and absorbing activities for people with dementia. As well as providing them with rewarding and satisfying Flow experiences, independent gaming would enable people with dementia to exert some control over their environment, for which they have fewer and fewer opportunities in the rest of their lives. Additionally, the finding that people with dementia can actually learn new information and activities, just not through explicit means, has implications beyond the development of games to the broader application of creating technological solutions to the difficulties faced by people with a dementia diagnosis.

These findings support the view that it is possible to develop cognitive prostheses to leverage and extend capacities and to develop systems that fit the human and machine components together in ways that synergistically exploit their respective strengths and mitigate weaknesses (Institute for Human and Machine Cognition, 2012). For people with dementia this means taking account of their working memory difficulties, which make it difficult for them to learn new information, while maximizing their unaffected abilities. Projects such as Age Invaders demonstrate the potential for developing systems that take account of specific user needs, enabling them to be maximally useful for people with cognitive impairments.

Fig. 10.1 The botanic garden

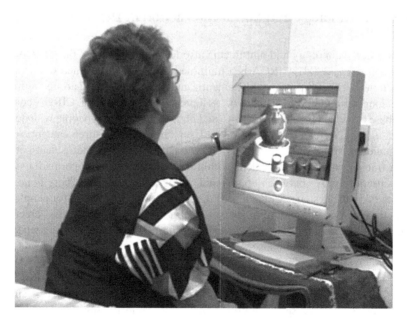

Fig. 10.2 Painting a pot

In addition to cognitive stimulation, technology has the potential to contribute to people with dementia living meaningful and enjoyable lives in a number of ways. For example, it is well established that musical memory, that is, memory for well-rehearsed music and songs, is well retained, and a number of projects have explored the potential of providing music to benefit people with dementia. The Picture Gramophone was an early small-scale project (Topo et al., 2004) to develop a music-based multimedia program to engage people with dementia. Evaluations with a small number of participants in several European countries were positive and were recently extended by Sixsmith, Orpwood, and Torrington (2010) in their development of a music player for people with dementia to access music for themselves, as opposed to having to rely on a caregiver. The preliminary results suggest that this may be of most benefit to people whose dementia has progressed to a point where they are living in nursing homes but there is further work to do.

Another small-scale pilot project explored the potential for enabling people with dementia to make music using a touch screen computer rather than just listening to it (Riley, Alm, & Newell, 2009). Typically, music groups and music therapy made available for people with dementia are conducted in groups using percussion instruments as these are most accessible for novices. In this project Riley and colleagues attempted to combine specific emotions with major and minor chords to provide an engaging experience for people with a dementia diagnosis. The initial findings suggested that the system was easy to use and engaging but requires further work to fully explore its potential.

10.15   Technology and Social Interaction in Dementia

Touch screen technology and multimedia also formed the basis of the CIRCA project (Astell et al., 2004). CIRCA is a multimedia computer designed to encourage interactions between a person with dementia and a caregiver based around reminiscing, as long-term memories tend to be well preserved in dementia. CIRCA contains a database of video clips, music, songs, and photographs, which people with dementia and a caregiver can explore together. Users choose from three categories (e.g. entertainment, sport, and schooldays) via a touch screen and then select between media (music, photographs, video) within that category. The hypermedia system allows users to move between text, sound, and graphics at will. There is no *right place* to be in the system (McKerlie & Preece, 1993), which is ideal for people with dementia who can enjoy current stimuli, such as a photograph on the screen, but have difficulty keeping topics of conversation in mind and remembering what they were previously talking about.

CIRCA was developed to take account of and respond to the profiles of spared and impaired cognitive processes of people with dementia (Astell et al., 2008). As such, CIRCA follows in the tradition of Augmented Cognition in that it aims to work with people's preserved abilities and skills while finding ways round their impaired ones. CIRCA was tested in one-to-one sessions and compared with traditional reminiscence activities, such as looking at picture cards or old artefects.

In comparison to traditional reminiscence, which was typically a question and answer session led by caregivers, CIRCA facilitated a shared activity that both parties could enjoy together. This appears to be due to the contents being used to stimulate reminiscing being novel to both parties, which allows them to explore it together, with neither having any advantage in knowing what will happen. This has the effect of restoring people with dementia to equal partners in the interaction. Often it is the people with dementia touching the screen to make choices and lead the interaction (Astell et al., 2010).

Besides direct benefits to people with dementia from using CIRCA, they also benefit from the enjoyment and satisfaction of caregivers who find CIRCA a supportive tool for interacting with people with dementia. Caregivers particularly enjoyed the ease with which people with dementia were able to use the system and the reminiscences that are sparked. CIRCA challenges the often very low expectations that formal and family caregivers have about what people with dementia are capable of. In post-session interviews, caregivers felt that they learned more about the person with dementia after using CIRCA and often saw them in a new light, as a person with a history they had just discovered. This has obvious potential benefits for the future of their relationships, which should in turn impact on the quality of care and quality of life of people with dementia.

10.16 Potential Future Developments of Technology for Dementia

Story Table™ by the de Waag Society is another technology project for people with dementia based on reminiscence. Like Eldergames, Story Table uses a tabletop presentation and contains short video clips of 5–15 min from the 1920s to 1950s. A study of 181 older people resident in care facilities in the Netherlands identified three functions of the system: fun, promotion of social contact, and eliciting storytelling (Knipscheer, Nieuwsteeg, & Oste, 2006). The potential for promoting intergenerational storytelling and other possible applications for older people with dementia in nursing homes is still being explored.

Tabletop touch screens are also being explored for providing access to art therapy for people with dementia living in care facilities (Hoey, Leuty, Zutis, & Mihailidis, 2010). This project is attempting to use technology to enable art therapists to work with a number of clients at one time and, like the making music project described above, works with preset activities, such as collage and painting, presented by touching the screen. As with the other projects currently under way, this work recognizes that people whose dementia has reached the more advance stages also deserve to have their need for engagement in creative and engaging activities met.

To date there has been less research than with older adults generally into potential applications of technology to meet the psychological needs of people living with a diagnosis of dementia. However, the findings discussed highlight and demonstrate not only the possibilities of designing for dementia but also the benefits of this. Tackling the cognitive and social difficulties of dementia and the tendency of caregivers and the environment to excessively disable them is to tackle a fundamental aspect of what it is to be human and to address how a disorder such as dementia challenges the very foundations of personhood.

In summary, technology can provide solutions to meeting the psychosocial needs of people with dementia, which is an aspect of care that is often overlooked in the pressure to keep people safe and secure. Projects such as LIM and CIRCA, which were designed to enhance the lives of people living with dementia and support their relationships with caregivers, also demonstrate the potential and importance of technology in responding to the ever-increasing numbers of older people affected by this progressive condition of aging.

10.17 Conclusions

To date, there has been fairly minimal exploration of technology to support people to live well as they age, beyond addressing very basic human needs. This is particularly so in respect of older people living with dementia (Sixsmith, 2006). This may reflect a lack of interest in meeting the higher-level needs of older people or, more charitably, recognition that this is a difficult task. However, as the population shifts inexorably towards a longer older age, the time for taking on this task, no matter how difficult, is at hand.

We must harness the potential of technology, not only to ensure people are safe and secure as they age but increasingly to ensure that their old age is happy and worth living. The evidence for creative applications of technology is growing, but the pace needs to increase substantially if we are to reap the benefits in the next 20 years. Positive psychology can provide a starting point and frame of reference for developing technology to bring people fun, meaning, and happiness. The challenge is to rethink our attitudes to aging and what people want as they age.

This requires detailed recognition of the complex needs of the heterogeneous population of people we class as old—that is, people over 65 years of age. Increasingly this group includes two generations of older people within one family—the old and the so-called old-old. The implications of this new but growing situation are only just being recognized but highlight the urgency of creating the next generation of novel technologies to enable people to live and age as well as possible.

References

Age Concern. (2008). *Undiagnosed, untreated, at risk: The experiences of older people with depression*. London: Age Concern.

Alzheimer's Association. (2010). Accessed September 30, 2010 from http://www.alz.org/index.asp

Anderson, D. N. (2001). Treating depression in old age: The reasons to be positive. *Age and Ageing, 30*, 13–17.

Astell, A. J. (2006). Personhood and technology in dementia. *Quality in Ageing, 7*(1), 15–25.

Astell, A. J. (2010). Developing computer games for people with dementia. *Gerontechnology, 9*(2), 89.

Astell, A. J., Alm, N., Gowans, G., Ellis, M. P., Dye, R., & Campbell, J. (2008). CIRCA: A communication prosthesis for dementia. In A. Mihailidas, L. Normie, H. Kautz, & J. Boger (Eds.), *Technology and aging*. Amsterdam: IOS Press.

Astell, A. J., Alm, N., Gowans, G., Ellis, M., Dye, R., Campbell, J., et al. (2009). Working with people with dementia to develop technology: The CIRCA and living in the moment projects. *Journal of Dementia Care, 17*, 36–38.

Astell, A. J., Ellis, M. P., Alm, N., Dye, R., Campbell, J., & Gowans, G. (2004). Facilitating communication in dementia with multimedia technology. *Brain and Language, 91*(1), 80–81.

Astell, A. J., Ellis, M. P., Bernardi, L., Alm, N., Dye, R., Gowans, G., et al. (2010). Using a touch screen computer to support relationships between people with dementia and caregivers. *Interacting with Computers, 22*, 267–275.

Astell, A. J., & Orpwood, R. D. (2010). Prompting to support independence in dementia. *Gerontechnology, 9*(2), 146.

Bail, K. D. (2003). Devices may be preferable to locked doors. *British Medical Journal, 326*(7383), 281.

Bialystok, E. (2006). Effect of bilingualism and computer video game experience on the Simon task. *Canadian Journal of Experimental Psychology, 60*(1), 68–79.

Brayne, C., Ince, P. G., Keage, H. D., McKeith, I. G., Matthews, F. E., Polvikoski, T., et al. (2010). Education, the brain and dementia: Neuroprotection or compensation? *Brain, 133*(8), 2210–2216.

Brunnström, H., Gustafson, I., Passant, U., & Englund, E. (2009). Prevalence of dementia subtypes: A 30-year retrospective survey of neuropathological reports. *Archives of Gerontology and Geriatrics, 49*(1), 146–149.

Coughlin, J. (2010). Disruptive Demographics. Accessed September 30, 2010 from http://www.disruptivedemographics.com/p/about-disruptive-demographics.html

Csikszentmihalyi, M. (1990). *FLOW: The psychology of optimal experience*. New York: Harper Row.

Gamberini, L., Alcaniz, M., Barresi, G., Fabregat, M., Ibanez, F., & Prontu, L. (2007). Cognition, technology and games for the elderly: An introduction to ELDERGAMES project. *PsychNology Journal, 4*(3), 285–308.

Gamberini, L., Martino, F., Seraglia, B., Spagnolli, A., Fabregat, M., Ibanez, F., et al. (2009). *Eldergames project: An innovative mixed reality table-top solution to preserve cognitive functions in elderly people*. IEEEExplore. Accessed September 30, 2010 from http://ieeexplore.ieee.org/stamp/stamp.jsp?arnumber=05090973

Graves, L., Ridgers, N., Williams, K., Stratton, G., Atkinson, G., & Cable, N. (2010). The physiological cost and enjoyment of Wii Fit in adolescents, young adults, and older adults. *Journal of Physical Activity and Health, 7*(3), 393–401.

Hoey, J., Leuty, V., Zutis, K., & Mihailidis, A. (2010). A tool to promote prolonged engagement in art therapy: Design and development from arts therapist requirements. In *Proceedings of 12th International ACM SIGACCESS Conference on Computers and Accessibility*, Orlando, FL.

Hofman, M., Hock, C., Kuller, A., & Müller-Spahn, F. (1996). Interactive computer-based cognitive training in patients with Alzheimer's disease. *Journal of Psychiatric Research, 30*(6), 493–501.

Hofman, M., Hock, C., & Müller-Spahn, F. (1996). Computer-based cognitive training in Alzheimer's disease patients. *Annals of the New York Academy of Sciences, 777*, 249–254.

Hofman, M., Ròsler, A., Schwarz, W., Müller-Spahn, F., Kräuchi, K., Hock, C., et al. (2003). Interactive computer-training as a therapeutic tool in Alzheimer's disease. *Comprehensive Psychiatry, 44*(3), 213–219.

Ijsselsteijn, W. A., Nap, H. H., de Kort, Y. A. W., & Poels, K. (2007). *Game design for elderly users* (Presented at Future Play 2007, Toronto, Canada). Pensacola, FL: Institute for Human and Machine Cognition (IHMC).

Institute for Human and Machine Cognition. (2012). *Website*. Retrieved from http://www.ihmc.us/aboutIHMC.php

Jimison, H., & Pavel, M. (2007). Integrating computer-based health coaching into elder home care. In A. Mihailidas, L. Normie, H. Kautz, & J. Boger (Eds.), *Technology and aging*. Fairfax, VA: Ios Press.

Jimison, H. B., Pavel, M., Bissell, P., & McKanna, J. (2007). A framework for cognitive monitoring using computer game interactions. *Studies in Health Technology and Informatics, 129*(Pt 2), 1073–1077.

Katzman, R. (1993). Education and the prevalence of dementia and Alzheimer's disease. *Neurology, 43*(11), 13–20.

Khoo, E. T., Merritt, T., & Cheok, A. D. (2009). Designing physical and social intergenerational family entertainment. *Interacting with Computers, 21*(1–2), 76–87.

Knipscheer, K., Nieuwsteeg, J., & Oste, J. (2006). Persuasive story table: Promoting exchange of life history stories among elderly in institutions. In W. IJsselsteijn et al. (Eds.), *Persuasive technology* (Lecture Notes in Computer Sciences, Vol. 3962, pp. 191–194). Berlin: Springer.

Maslow, A. H. (1943). A theory of human motivation. *Psychological Review, 50*, 370–396.

Mate-Kole, C. C., Fellow, R. P., Said, P. C., McDOugal, J., Catayong, K., Dang, V., et al. (2007). Use of computer assisted and interactive cognitive training programmes with moderate to severely demented individuals: A preliminary study. *Aging & Mental Health, 11*(5), 485–495.

McKerlie, D., & Preece, J. (1993). The hype and the media: Issues concerned with designing hypermedia. *Journal of Microcomputer Applications, 16*(1), 33–47.

National Institute of Mental Health. (2007). *Suicide in the U.S.: Statistics and prevention* (NIH Publication No. 06-4594). Bethesda, MD: Author.

Pigford, T., & Andrews, A. W. (2010). Feasibility and benefit of using the Nintendo Wii Fit for balance rehabilitation in an elderly patient experiencing recurrent falls. *Journal of Student Physical Therapy Research, 2*(1), 12–20.

Richards, M., & Deary, I. J. (2005). A life course approach to cognitive reserve: A model for cognitive aging and development? *Annals of Neurology, 58*(4), 617–622.

Riley, P., Alm, N., & Newell, A. (2009). An interactive tool to promote musical creativity in people with dementia. *Computers in Human Behavior, 25*(3), 599–608.

Salzman, C. (1995). Medication compliance in the elderly. *Journal of Clinical Psychiatry, 56*(Suppl 1), 18–22. Discussion 23.

Seligman, M. E. P., & Csikszentmihalyi, M. (2000). Positive psychology: An introduction. *American Psychologist, 55*, 1–14.

Sixsmith, A. (2006). New technologies to support independent living and quality of life for people with dementia. *Alzheimer's Care Quarterly, 7*(3), 194–202.

Sixsmith, A. J., Orpwood, R. D., & Torrington, J. M. (2010). Developing a music player for people with dementia. *Gerontechnology, 9*(3), 421–427.

Stern, Y. (2006). Cognitive reserve and Alzheimer's disease. *Alzheimer's Disease and Associated Disorders, 20*(2), 112–117.

Stern, Y., Alexander, G. E., Prohovnik, I., & Mayeux, R. (1992). Inverse relationship between education and parietotemporal perfusion deficit in Alzheimer's disease. *Annals of Neurology, 32*, 371–375.

Takatalo, J., Nyman, G., & Laaksonen, L. (2008). Components of human experience in virtual environments. *Computers in Human Behavior, 24*(1), 1–15.

Topo, P., Mäki, O., Saarikalle, K., Clarke, N., Begley, E., Cahill, S., et al. (2004). Assessment of a music-based multimedia program for people with dementia. *Dementia, 3*(3), 331–350.

Vaillant, G. E., & Mukamal, K. (2001). Successful aging. *American Journal of Psychiatry, 158*, 839–847.

Vorghese, J., Lipton, R. B., Katz, M. J., Hall, C. B., Derby, C. A., Kuslansky, G., et al. (2003). Leisure activities and the risk of dementia in the elderly. *New England Journal of Medicine, 348*, 2508–2516.

Whitcomb, G. R. (1990, September). Computer games for the elderly. In *Proceedings of the Conference on Computers and the Quality of Life* (pp. 112–115), Washington DC, USA.

Wimo, A., & Prince, M. (2010). *World Alzheimer report 2010: The global economic impact of dementia*. London: Alzheimer's Disease International.

World Health Organization. (1948). *Preamble to the Constitution of the World Health Organization as adopted by the International Health Conference*, New York, 19 June–22 July 1946; signed on 22 July 1946 by the representatives of 61 States (Official Records of the World Health Organization, no. 2, p. 100) and entered into force on 7 April 1948.

Zelinksi, E. M., & Reyes, R. (2009). Cognitive benefits of computer games for older adults. *Gerontechnology, 8*(4), 220–235.

Chapter 11
Videoconferencing and Social Engagement for Older Adults

Robert Beringer and Andrew Sixsmith

11.1 Introduction

When we think about *activity* in later life, our first thoughts often take us towards the idea of physical activity and the person's ability to remain fit, physically robust, and mobile. The World Health Organization (WHO) has taken a broader view and defined *active aging* as a state which "allows people to realize their potential for physical, social, and mental well-being throughout the life course and to participate in society" (WHO, 2012, p. 1). In this chapter our focus will be on the social dimension of this construct and specifically the potential role that videoconferencing may have in assisting older adults to remain socially active and engaged. We begin by reviewing literature that highlights the positive health and benefits that accrue to older adults who remain socially active. This is followed by the results of qualitative research that illustrates the potential value of videoconferencing technology in remaining socially connected.

11.2 The Social Dimension of Active Aging

The WHO (2012) defines active aging in very broad terms, including continued participation in all cultural, social, economic, spiritual, and civic affairs. This social component, namely, our continued interaction with family and friends, or what may be termed one's *social support network* is the focus of this chapter.

R. Beringer (✉)
Vibrant Living and Lifestyle Systems INC.,
124 Sandpiper Place, Salt Spring Island, BC, Canada V8K2W5
e-mail: myresearchrms@mac.com

A. Sixsmith
Gerontology Research Centre, Simon Fraser University, Harbour Centre,
#2800-515 West Hastings Street, Vancouver, BC, Canada V6B 5K3
e-mail: Sixsmith@sfu.ca

A. Sixsmith and G. Gutman (eds.), *Technologies for Active Aging*,
International Perspectives on Aging, DOI 10.1007/978-1-4419-8348-0_11,
© Springer Science+Business Media New York 2013

Maintaining social connections and a sense of support is far more than just a pleasant way to live one's life; it is a key component to remaining healthy as we age. While it is beyond the scope of this chapter to provide a full review of the health benefits of social support, a number of key components are worth discussing in relation to active aging. First, on a physiological level, social support has been found to positively impact physical well-being. For example, measures of physical health are positively influenced by social connectedness, including blood pressure (Hawkley, Masi, Berry, & Cacioppo, 2006) and lower cortisol levels (Turner-Cobb, Sephton, Koopman, Blake-Mortimer, & Spiegel, 2000). Further, recent studies suggest that social support is related to better immune function (Dixon et al., 2001; Esterling, Kiecolt-Glaser, & Glaser, 1996; Lutgendorf et al., 2005; Miyazaki et al., 2005). Second, on a psychological level, both isolation and inadequate social support have been identified as risk factors for depression in older adults (Bruce, 2002; Chi & Chou, 2001; George, Blazer, Hughes, & Fowler, 1989). Additionally, Vanderhorst and McLaren (2005) found that older adults with fewer social resources had an increased risk for depression and suicide ideation. Given that the population is aging worldwide (United Nations Population Division, 2002) and older adults are at higher risk for depression than the general population (National Institute of Mental Health, 2010), it is important that we seek out strategies to assist older adults in maintaining social support. One such strategy that has recently emerged as a possible technology-based solution is the use of videoconferencing as a tool to enhance social connectivity in older adults.

11.3   Videoconferencing

According to Nefsis (2008) "when video conferencing was first introduced, with a grand introduction at the 1964 World's Fair held in New York, it had a futuristic quality that no one could possibly dream would take the place of the standard telephone" (p. 1). Like many new technologies videoconferencing was expensive and went through many iterations until the present day where it is fast becoming commonplace. A major factor in the current expansion of videoconferencing has been the availability of high-speed Internet at reasonable cost, something that has only been for around a decade. Videoconferencing has the potential to impact both community-dwelling older adults and those who live in long-term care facilities. Interestingly, to date the research concerned with community-dwelling older adults has not been focused on maintaining social connectivity. Instead, disease self-management (Demiris, Speedie, & Finkelstein, 2001), rehabilitation support (Hauber & Jones, 2002), and medical applications such as vital signs monitoring and consultation between patient and healthcare professional (DelliFrane & Dansky, 2008; Koch, 2005) have been the focus. Turning to older adults in long-term care, the early results on videoconferencing and social support have been promising. One very early study by Mickus and Luz (2002) reported that videophones had the potential to reduce social isolation in frail nursing home residents. It is worthy to note that much of the qualitative data obtained from this research was concerned with technical issues surrounding the videophones that were being used. Demiris et al. (2008) in a study where technical issues were less of a concern reported that videophone contact could play a role in reducing both social isolation and loneliness of residents in long-term care. In similar research, Hsiu-Hsin, Yun-Fang, Hsui-Hung, Yue-Cune, and Hao (2010) found that videoconferencing alleviated feelings of loneliness and improved depressive status after 3 months in elderly nursing home residents. While such electronically mediated communication remains second best to an in-person visit (Tsai & Tsai, 2010), residents enjoyed being able to see into the homes of their family members. Research reported by Sixsmith (2009) describes the use of *A Window on the World Technology* that continuously streamed video from local neighborhood onto dedicated screens in nursing homes and the homes of community-dwelling older people. The sense of connectedness to the outside world or with previous familiar environments afforded by these simple video-based technologies has strong potential to alleviate the social isolation and disconnection that many older people experience as a defining feature of their everyday life.

The research that will be presented in this chapter emerged as an important theme within a wider research project on the use of information and communication technologies to enhance the lives of older people. The findings described below highlight the potential role of videoconferencing and suggest that video-based technologies and applications should be explored further as a tool to reduce social isolation and positively impact the lives of community-dwelling older adults.

11.4 Design and Original Research Question

This chapter reports on research that was part of a wider study of older people's attitudes towards Ambient Assisted Living (AAL) systems. AAL systems may be described as a combination of stand-alone assistive devices, smart home innovations, and telehealth technology to assist a person to live independently at home and remain connected. Much of the AAL research to date had been concerned with technical issues and/or acceptance of this technology, and in contrast we explored the potential for this technology to impact users' *meaning of home* (Sixsmith, 1990). As a visualization tool, we showed the film *Imagine the Future of Aging* to a number of older people provided for use in this research by the Center for Aging Services Technology (CAST, 2005). The film runs 9 min and 46 s and depicts the following features of an AAL system:

- Medication reminding
- Vital signs monitoring
- In-home monitoring of the activities of daily living via sensor technologies
- Displays that provide both professional and informal caregivers (e.g., family) with access to data collected
- An electronic health record

- Cognitive function monitoring by way of an assessment tool embedded in a solitaire game
- Social interaction with an online community of friends
- Videoconferencing with caregivers and healthcare professionals
- Social connectivity via videoconferencing with distant family members

The film portrays the AAL system being used by a man called Ernesto.

Since 2005 when the film was produced, videoconferencing has become more widely used. As a result, at the time we collected our data in the summer of 2010, we were able to explore more specifically whether videoconferencing could help community-dwelling older adults remain socially active as they age.

11.5 Approach

A qualitative approach was employed in this study and was deemed appropriate due to the exploratory nature of this research (Creswell, 2003). The process was iterative, meaning data were gathered and analysis took place concurrently over the course of the study (Bryman & Teevan, 2005). The data were collected by way of semi-structured interviews with research participants, using a set of specific questions, while maintaining a flexible approach outlook with regard to how the interview is to proceed. The rationale behind this approach is derived from the idea that excessive structure may hinder "genuine access to the worldviews" (p. 186) of research participants. Further, this open-ended method of data collection is considered to generate rich and varied information (Sixsmith, 1986).

One limitation of the method was that participants would not be able to truly visualize AAL (and videoconferencing) in their own homes and be able to comment authentically on such a system after merely viewing a film. Additionally, if participants did not *identify* with Ernesto, they may then have difficulty in making empathetic responses. To deal with these issues, participants were (if they could not see themselves using the technologies) asked to put themselves into the place of Ernesto. Interestingly, they often not only envisioned the system through Ernesto's eyes but responded using a first-person narrative.

Approval of the Simon Fraser University Ethics Review Board was obtained prior to commencement of this research (Simon Fraser, 2006), and pseudonyms have been used to protect the confidentiality of research participants.

11.6 Sample

The sample comprised 27 community-dwelling individuals aged 60 years and over, with at least one self-reported chronic condition and/or mobility restriction. A self-reported chronic condition was operationalized in accordance with the Canadian

Community Health Survey (Gilmour & Park, 2006) where participants must have been suffering from at least one of 20 chronic conditions, including but not limited to arthritis, cataracts, heart disease, asthma, diabetes, bronchitis, cancer, and effects of a stroke. A mobility restriction was operationalized as an individual who uses a cane, wheelchair, walker, or scooter to assist in mobility. We sought out the sample described above because it was felt that individuals with such conditions and/or restrictions would be more likely to use an AAL system if it were available. A combination of convenience and snowball sampling was used to recruit research participants in and around Vancouver, British Columbia, and San Francisco, California. Additional participants were recruited using snowball sampling where participants who had completed the study provided suggestions for where to find new recruits (Bryman & Teevan, 2005). Alternative sampling techniques may have been employed, but the fact remains that it is difficult to convince older adults to participate in research studies, and a selection bias towards overly healthy samples is common (Spirduso, Francis, & MacRae, 2005). In the present study, the use of convenience and snowball sampling resulted in access to participants who were more likely to benefit from technologies such as videoconferencing to remain socially connected.

Approximately half of the sample (12) were from British Columbia and 15 were from California. All lived alone in their own homes. Seventeen were female, and ten were male. They ranged in age from 60 to 95 (mean age = 77.7 year). In terms of health and mobility characteristics, the group had a mean of 4.0 chronic conditions and 48.2 % of participants ($n=13$) relied on an assistive device to function within the home. In terms of self-reported health, 14.8 % of participants ($n=4$) rated their health as excellent, 40.7 % ($n=11$) as very good, 25.9 % ($n=7$) as good, 4.8 % ($n=4$) as fair, and 3.7 % ($n=1$) as poor. Just under one quarter of the participants ($n=5$) had experienced a fall in their home. One of these falls was serious enough to result in broken bones, and two participants reported falling in the evening and having to remain on the floor overnight until help arrived. The majority of participants (77.8 %) admitted to experiencing pain on a regular basis with 14.8 % ($n=4$) reporting very mild pain, 33.3 % ($n=9$) mild pain, 22.2 % ($n=6$) moderate pain, and 7.4 % ($n=2$) severe pain. Twelve (44.4 %) reported being hospitalized in the year prior to this study. The mean length of stay for these hospitalizations was 7.6 days (range 0–60 days).

11.7   Data Analysis

Interviews were transcribed verbatim and analyzed using NVivo 8, a software package designed to classify, sort, and arrange qualitative research data (QSR International, 2009). Using NVivo 8, the coding of data was completed through a line-by-line analysis and the development of nodes. The nodes represent salient behaviors and concepts of interest or themes (Bryman & Teevan, 2005). The theme of videoconferencing that resulted from this process is the focus of this chapter.

11.8   Results

Twenty-five of the 27 participants made at least one comment on videoconferencing technology. A total of 43 positive comments were logged as well as seven that were negative or cautionary in nature. Nine of the 27 (33 %) participants had actually used videoconferencing themselves. Interestingly, the age range of these nine participants was 61–88 years, and the mean age of those who had used videoconferencing was 72.6 year only slightly below the mean age of all participants (77.7 year) in the study.

The comments on videoconferencing were generally very positive. For example, Meg (aged 71) noted:

> I think I could see that being beneficial, say for instance in my case, both of my family members are not living in the same town as me, so in that respect I think it would put their minds at ease and mine too.

Gina (aged 79) who had never used the technology stated:

> Oh it would be wonderful because that way you're in touch with family, like the family that I have left are all in the east, and they, I'll get phone calls and stuff from them but it's not quite the same as that (referring to videoconferencing).

Videoconferencing was seen to foster a stronger sense of proximity and connectedness compared to conventional telephones. Elizabeth (aged 75) was very emphatic on this point:

> It (videoconferencing) gives you a feel of more, being closer, you know, I mean you talk on the phone and it's fine you know, you have the contact, but you actually see them, you see the kids playing in the back and they say 'hi grandma' and it for me, it gives me a more closer feel to the family.

Similar commentary was echoed by Carla (aged 88) who when asked what it has been like to use videoconferencing stated:

> It's wonderful because I have a lot of nieces and nephews back east and with new babies and stuff and I got to see them… you feel right close to them, you know.

The following passage from the interview with research participant Anita (aged 81) highlights the strong sense of emotional connectedness afforded by video-based communication.

(Interviewer questions in **bold print**):

> **Do you have any people who live far away that you talk to on the telephone?**
> "We had nieces and nephews, but we had them pulled up on the computer, they were in Denver."
>
> **So you've seen some people on the computer using videoconferencing?**
> "Yes."
>
> **So tell me how that felt, when you saw somebody you were talking with, out of state**

"Oh nice, I mean we all had tears, you know because it was a holiday and they couldn't come home, yeah, oh it's great."

Do you think you would have had tears if you had chatted with them on the phone?
"No."

In contrast, some of those who had not used videoconferencing were more likely to comment in the manner of Jane (aged 91) who told us:

I think the phones are good enough, and I, my daughter and I talk every night, and that works well for me, but for other people who don't have family close, then that's a different thing.

In addition to this more social aspect of videoconferencing, we also probed as to other possible benefits of this technology. Here a subtheme emerged which suggested that videoconferencing may be used as a tool to monitor a person's health. For instance, Albert (aged 88) envisioned that:

I think it's excellent to be able to see them… perhaps if the person was ill, or recovering from an illness then I would like to see how they looked.

Following on this theme, Donna (aged 62) noted:

Well there's a couple of things, you can see their mood and there, like if he talks to his son, you can tell if his son is in a good mood or a bad mood, you can see there well-being.

While most of the participants were positive about the use of videoconferencing, we collected a number of responses that were either negative or at least cautionary in nature. More than one participant commented on the disadvantages of *being seen,* and Alex (aged 66) notes that:

I am positive it would (change the way I feel about my home) because I would have to dress up for all occasions, I couldn't just walk disheveled and unshaven and what not because I would be watched, so it changes your behavior a lot.

In his comment, Alex felt the telephone, in which one is heard but not seen, was a more relaxing way to carry out day-to-day communication.

Most of the negative comments revolved around whether videoconferencing technology could actually lead to increased isolation and particularly that it could replace in-person contact. Debra (aged 77), for example, suggested:

I think it could be used as an excuse not to go and see each other because that interaction only lasts a few minutes and could easily be substituted for real visiting.

In like manner, Elizabeth cautions that:

In my opinion there should still be some human interaction, yes. I mean I agree with all those devices I really do but I think you shouldn't lose completely the personal contact even if it's just a social worker, whatever, caretaker, somebody.

In a very interesting contrast, Sarah (aged 78) noted that:

Because you wouldn't need them to come and visit you as much if you could just plug into that and see them, know that they're well, and they can do the same for you.

Another theme that emerged concerned the amount of learning required to use videoconferencing. For example, Andy (aged 81) when asked if he would be interested in using videoconferencing stated:

Well, I would think it could be (fun to use) depending on how difficult it is to figure out

Similarly, Marcus (aged 74) commented:

I think the problem is the process of, with me it's the process of knowing a computer language and looking at a window and it's giving you options, do you want to do this, and yes, no, and half the time it's hard to know what they're really getting at, and older people like me who have never studied computers find the computer language difficult and even scrolling down or knowing that you have to scroll down, it's very hard.

In some ways, Marcus's words reflect the prevalent view that older adults are technophobic, with generally negative attitudes towards using new types of technologies. However, it should be emphasized that these were very much the minority among our participants (X) compared to those who expressed positive views (Y) who, after learning a given technological task, described benefits in terms of feeling connected with family and friends.

11.9   Discussion

Jacques Ellul (1962) wrote over 50 years ago:

The further we advance into the technological society, the more convinced we become that, in any sphere whatever, there are nothing but technical problems. We conceive all problems in their technical aspect, and think that solutions to them can only appear by means of further perfecting techniques. (p. 414)

Ellul's comment is relevant to the present discussion because currently much of the research on technology and older adults focuses on what Sixsmith et al. (2010) have coined *problematization of aging, where technological solutions are fixated on an overwhelmingly negative agenda of dependency and ill health*. As an alternative, instead of identifying social isolation or decreased social interaction as problems and applying videoconferencing as a solution, we suggest that it may be better to think about using this technology further upstream. Much of the data we have gathered in this study suggests that videoconferencing may be helpful in the sense of promoting social connectedness amongst older adults and thereby potentially encouraging and facilitating social connectedness. The idea of active aging includes the idea that people can maintain social connections and well-being throughout the life course (WHO, 2012). Our participants who had used videoconferencing stated that it made them feel as if they had actually *visited* distant grandchildren, i.e., that it enhanced their *social connectivity*. The issue here goes well beyond the usual engineering perspective that focuses on developing specific solutions to identified problems. Indeed the issue here is not the development of new technologies (the technologies described are readily available to the general public) but that older people are often marginalized by technological change, contributing to social disconnection as the new technology restructures the way we interact as social beings.

Many older people are for various reasons (money, skills, attitudes) excluded emerging technologies, yet the research here suggests both a desire of many older people to be able to use and benefit from new communication technologies. Indeed it could be argued that the creation of a class of technologies, such as AAL systems that are specifically targeted at older people, may ironically serve to maintain social divisions within the technological sphere, especially where technologies are labeled as *solutions* to *problems* of old age, suggesting a highly medicalized view of the nature of old age. The social construction of technology as a solution reflects a discourse of dependency and passivity, as opposed to a discourse of technology as a means for social participation and active aging.

While the emphasis of this chapter has been very much on the value of videoconferencing to enhance the social well-being of older people, it is important to highlight some of the potential disbenefits of technology-based communications and its potential to impinge upon the intimate sphere of the home and to replace direct human contact with electronically mediated social interaction. Goffman has asked the question "of the more peaceful and secure… what steps would be necessary to transform it into something that was deeply unsettling" (p. 61, cited in Kim, 2003). In the case of videoconferencing the deeply unsettling view is of an older adult sitting alone, void of human contact, connected to the outside world only by way of a two-way screen. This concern was expressed by participants in our study and echoes concerns of participants in a study by Lorenzen-Huber, Boutain, Camp, Shankar, and Connelly (2010). It should be noted however that little research has investigated this potential darker side of videoconferencing technology. While our research highlights the potential of technology in remaining socially active and connected, future research should be directed towards exploring in greater depth the way that technology generally and videoconferencing specifically is integrated within the everyday lives of older people. This requires us to shift in focus from a medical engineering agenda to a social research agenda that examines how technology transforms social relationships and the conditions of everyday life.

References

Bruce, M. (2002). Psychological risk factors for depressive disorders in late life. *Biological Psychiatry, 52*, 175–184.

Bryman, A., & Teevan, J. (2005). *Social research methods*. Don Mills: Oxford University Press (Canadian Ed.).

Center for Aging Service Technology. (2005). *Imagine the future of aging*. Accessed May 5, 2012 from http://www.leadingage.org/Imagine-the-Future-of-Aging.aspx

Chi, I., & Chou, K. (2001). Social support and depression among elderly Chinese people in Hong Kong. *International Journal of Aging & Human Development, 52*, 231–252.

Creswell, J. (2003). *Research design: Qualitative, quantitative, and mixed methods approaches* (2nd ed.). Thousand Oaks, CA: Sage.

DelliFrane, J., & Dansky, K. (2008). Home-based telehealth: A review and meta-analysis. *Journal of Telemedicine and Telecare, 14*, 62–66.

Demiris, G., Parker-Oliver, D. R., Hensel, B., Dickey, G., Rantz, M., & Skubic, M. (2008). Use of videophones for distant caregiving: An enriching experience for families and residents in long-term care. *Journal of Gerontological Nursing 34*(7), 50–55. Accessed April 26, 2012 from http://eldertech.missouri.edu/files/Papers/Demiris/Use_of_Videophones_for_distant_caregiving.pdf

Demiris, G., Speedie, S. M., & Finkelstein, S. (2001). Change of patients' perceptions of TeleHomeCare. *Telemedicine Journal and e-Health, 7*, 241–248.

Dixon, D., Kilbourn, K., Cruess, S., Klimas, N., Fletcher, M. A., Ironson, G., et al. (2001). Social support mediates the relationship between loneliness and Human Herpesvirus-Type 6 (HHV-6) antibody titers in HIV + gay men following hurricane andrew. *Journal of Applied Social Psychology, 31*, 1111–1132.

Ellul, J. (1962). The technological order. *Technology and Culture, 3*, 394–422.

Esterling, B. A., Kiecolt-Glaser, J. K., & Glaser, R. (1996). Psychosocial modulation of cytokine-induced natural killer cell activity in older adults. *Psychosomatic Medicine, 58*, 264–272.

George, L., Blazer, D., Hughes, D., & Fowler, N. (1989). Social support and the outcome of major depression. *The British Journal of Psychiatry, 154*, 478–485.

Gilmour, H., & Park, J. (2006). Dependency, chronic conditions and pain in seniors. *Health Reports Supplement, 8*, 33–45.

Hauber, R. P., & Jones, M. L. (2002). Telerehabilitation support for families at home caring for individuals in prolonged states of reduced consciousness. *Journal of Head Trauma Rehabilitation, 17*, 535–541.

Hawkley, L. C., Masi, C. M., Berry, J. D., & Cacioppo, J. T. (2006). Loneliness is a unique predictor of age-related differences in systolic blood pressure. *Psychology and Aging, 21*(1), 152–164.

Hsiu-Hsin, T., Yun-Fang, T., Hsui-Hung, W., Yue-Cune, C., & Hao, H. C. (2010). Videoconferencing program enhances social support, loneliness, and depressive status of elderly nursing home residents. *Aging & Mental Health, 14*(8), 947–954.

Kim, K. (2003). *Order and agency in modernity.* Albany: State University of New York Press.

Koch, S. (2005). Home telehealth: Current state and future trends. *International Journal of Medical Informatics, 75*, 565–576.

Lorenzen-Huber, L., Boutain, M., Camp, L. J., Shankar, K., & Connelly, K. H. (2010). *Privacy, technology, and aging: A proposed framework.* Retrieved May 8, 2012 from http://www.cs.indiana.edu/~connelly/Papers/J8_PrivacyFramework.pdf

Lutgendorf, S. K., Sood, A. K., Anderson, B., McGinn, S., Maiseri, H., Dao, M., et al. (2005). Social support, psychological distress, and natural killer cell activity in ovarian cancer. *Journal of Clinical Oncology, 23*, 7105–7113.

Mickus, M. A., & Luz, C. C. (2002). Televisits: Sustaining long distance family relationships among institutionalized elders through technology. *Aging & Mental Health, 6*(4), 387–396.

Miyazaki, T., Ishikawa, T., Nakata, A., Sakurai, T., Miki, A., Kawakami, N., et al. (2005). Association between perceived social support and Th1 dominance. *Biological Psychology, 70*, 30–37.

National Institute of Mental Health. (2010). *Older adults: Depression and suicide facts* (Fact Sheet). Accessed April 26, 2012 from http://www.nimh.nih.gov/health/publications/older-adults-depression-and-suicide-facts-fact-sheet/index.shtml

Nefsis. (2008). *Webpage: Videoconferencing timeline.* Accessed April 26, 2012 from http://www.nefsis.com/Best-Video-Conferencing-Software/video-conferencing-history.html

QSR International. (2009). *NVivo 8 website.* Retrieved May 2, 2009 from http://www.qsrinternational.com/products_nvivo.aspx

Simon Fraser University. (2006). *University research ethics review.* Retrieved March 18, 2007 from http://www.sfu.ca/policies/research/r20-01.htm

Sixsmith, J. (1986). The meaning of home: An exploratory study of environmental experience. *Journal of Environmental Psychology, 6*, 281–298 [Electronic version].

Sixsmith, A. (1990). The meaning and experience of 'home' in later life. In B. Bytheway & J. Johnson (Eds.), *Welfare and the ageing experience* (pp. 172–192). Aldershot: Avebury.

Sixsmith, A. (2009). *Seniors in transition*. Retrieved from http://gordonsestateservices.wordpress.com/2009/03/05/smart-houses-may-keep-seniors-at-home-longer/

Sixsmith, A., Mueller, S., Lull, F., Klein, M., Bierhoff, I., Deleaney, S., et al. (2010). A user-driven approach to developing ambient assisted living systems for older people: The SOPRANO Project. In J. Soar, R. Swindell, & P. Tsang (Eds.), *Intelligent technologies for bridging the grey digital divide* (pp. 30–45). Hershey, PA: IGI Global.

Spirduso, W., Francis, K., & MacRae, P. (2005). *Physical dimensions of aging* (2nd ed.). Champaign, IL: Human Kinetics.

Tsai, H., & Tsai, Y. (2010). Older nursing home residents' experiences with videoconferencing to communicate with family members. *Journal of Clinical Nursing, 19*, 1538–1543.

Turner-Cobb, J. M., Sephton, S. E., Koopman, C., Blake-Mortimer, J., & Spiegel, D. (2000). Social support and salivary cortisol in women with metastatic breast cancer. *Psychosomatic Medicine, 62*, 337–345.

United Nations Population Division. (2002). *World population aging: 1950–2050*. Accessed April 26, 2012 from http://www.un.org/esa/population/publications/worldageing19502050/

Vanderhorst, R. K., & McLaren, S. (2005). Social relationships as predictors of depression and suicidal ideation in older adults. *Aging & Mental Health, 9*, 517–525.

World Health Organization. (2012). *Webpage: What is Active Aging?* Accessed April 26, 2012 from http://www.who.int/ageing/active_ageing/en/index.html

Chapter 12
International Initiatives in Technology and Aging

Andrew Sixsmith, Maria Carrillo, David Phillips,
Peter Lansley, and Ryan Woolrych

12.1 Introduction

This chapter provides an overview of some of the initiatives and activities in the area of gerontechnology that have emerged in different parts of the world. While it is not possible to provide a comprehensive and exhaustive review of the world scene, the aim is to highlight selected research in different geographical regions, specifically Europe, North America, and Asia-Pacific, and research at different levels: national, regional, and international. The chapter begins with a description of an international program, Everyday Technologies for Alzheimer's Care (ETAC), funded through a partnership between the Alzheimer's Association and the INTEL Corporation and

A. Sixsmith (✉)
Gerontology Research Centre, Simon Fraser University, Harbour Centre,
#2800-515 West Hastings Street, Vancouver, BC, Canada V6B 5K3
e-mail: Sixsmith@sfu.ca

M. Carrillo, Ph.D.
Medical & Scientific Relations, Alzheimer's Association, National Office,
225 N. Michigan Ave., Suite 1700, Chicago, IL 60601-7633, USA
e-mail: Maria.Carrillo@alz.org

D. Phillips
Lingnan University, Lingnan, Hong Kong
e-mail: phillips@ln.edu.hk

P. Lansley, B.Sc., M.Sc., Ph.D., M.C.I.O.B., F.C.O.T.
KT-EQUAL – Knowledge Transfer for Extending Quality Life, School of Construction
Management and Engineering, University of Reading, Whiteknights, URS Building,
PO Box 219, Reading RG6 6AW, UK
e-mail: p.r.lansley@reading.ac.uk

R. Woolrych
Gerontology Research Centre, Simon Fraser University, Harbour Centre, 515 West Hastings
Street, Vancouver, BC, Canada V6B 5K3
e-mail: ryan_woolrych@sfu.ca

A. Sixsmith and G. Gutman (eds.), *Technologies for Active Aging*,
International Perspectives on Aging, DOI 10.1007/978-1-4419-8348-0_12,
© Springer Science+Business Media New York 2013

designed to develop technological solutions for people with Alzheimer's disease (AD) and their caregivers and families. The second section focuses on emerging research on technology and aging in the Asia-Pacific region. The third section focuses on national-level programs and describes work being carried out in the UK funded by government agencies such as the Environment and Physical Sciences Research Council (EPSRC). The fourth section describes the very extensive program of work being funded by the European Union under the Ambient Assisted Living Joint Program. The chapter concludes by identifying some of the key common themes within the various research programs described and discusses the drivers, emerging directions, and opportunities in a rapidly expanding global research landscape.

12.2   Everyday Technologies for Alzheimer's Care

In 2003, the Alzheimer's Association and Intel Corporation launched a program to develop technological solutions for AD patients, caregivers, and families, specifically to help alleviate the burden on caregivers, improve patients' quality of life and the quality of care they receive, extend patient independence, and slow the onset of symptoms by keeping older adults cognitively active. Since that time, the Everyday Technologies for Alzheimer's Care (ETAC) initiative has disbursed nearly $5 million to some 26 research projects throughout the world. These projects include a diverse array of technologies and address four overlapping issues important in the care of individuals with AD: diagnosis, monitoring, treatment, and assistance. Many of these technologies are in the prototype stage and are being tested in real-world environments.

12.2.1   Diagnosis

Standard tests of cognition have been adapted for computer administration, which offers numerous advantages in terms of accessibility, standardization, precision, and accuracy. Computer versions of visuospatial tests may be particularly sensitive, since computers can render three-dimensional shapes and manipulate these shapes in space in order to bring out subtleties that are not possible to represent on paper. Virtual reality, an advanced form of human–computer interface that allows individuals to become immersed in simulated environments, takes this even further. For example, a 15-min virtual environment-based neuropsychological test battery delivered within the context of a virtual city has been developed and shown to correlate with standard neuropsychological testing (Parsons, Silva, Pair, & Rizzo, 2008). Another new technology uses wearable sensors to record movement and speech as a means of detecting cognitive decline and assessing its progression. Additionally, technologies such as the electroencephalograph and the magneto-encephalograph are providing a wealth of information about the different neural networks that are active or inactive during various brain tasks (Dale, Simpson, Foxe, Luks, & Worden, 2008). These technologies could be used not only for diagnostic purposes but also to provide feedback for individuals getting cognitive training and to monitor the effects of other treatments.

12.2.2 Monitoring

Technologies that monitor behavior or activity remotely through the use of sensors, data and video recorders, and wireless telecommunication devices may enable individuals to maintain their independence even as their memory and physical functioning declines or they begin to exhibit challenging behaviors. For example, falls are a major concern for older adults living alone. With a grant from the ETAC program, colleagues at the University of Missouri developed a *smart* carpet containing an array of pressure sensors that electronically detects falls and transmits an alert to a caregiver (Tyrer, Aud, Alexander, Skubic, & Rantz, 2007; Tyrer et al., 2006). Similar technology might be used to monitor changes in walking speed or gait and thus could be used to detect a decline in physical ability. Other technologies have been developed for more routine monitoring of individuals living alone to ensure safety and monitoring of activities of daily living (ADL) such as eating, medication use, and bathing. In the Automated Technology for Elder Assessment Safety in the Environment (AT EASE) pilot study, motion sensors and sensors placed on household hardware such as doors, medicine cabinets, refrigerators, faucets, and toilets allowed caregivers, family members, and facility staff to monitor a variety of activities using a web interface that allowed them to log on as frequently as desired (Mahoney, Mahoney, & Liss, 2009).

12.2.3 Treatment

Everyday technologies can play an important therapeutic role in individuals with cognitive impairment. Therapeutic applications include cognitive training, personal exercise training, reminiscence therapy, and art and music therapy. Reminiscence therapy addresses the loss of identity experienced by many people with dementia by creating multimedia biographies that help individuals remember their past (Massimi et al., 2008). These biographies may also provide solace to family members, reminding them of what their loved one was like before the disease progressed. The efficacy of reminiscence therapy varies over the course of the disease, and it may be less useful in advanced cases where the affected individual may be unable to participate in developing the biography. A variation on the remembering theme is the SenseCam, a small camera with a wide-angle (fish-eye) lens that is worn throughout the day, recording 1,000–1,500 snapshots of a person's daily activities (Bas,

Erdogmus, Ozertem, & Pavel, 2008). Playback of the photos at the end of the day helps users remember what they have experienced (Hayes et al., 2008). Music and art have also attracted more attention recently as being therapeutic for people with dementia, and again, technology is helping make these types of therapy more available by allowing people to engage in creative pursuits despite language, visuo-perceptual, and physical deficits. For example, Hyperscore, a music-composition computer program, allows people with no music education to compose their own music (Sorrell & Sorrell, 2008). Music interfaces are also being used to enable early detection of AD with auditory tasks that target neurological structures affected at the earliest stages of the disease.

12.2.4   Assistance

Monitoring technologies with feedback can assist elders in performing ADLs or call for caregiver assistance if necessary. One such system called the COACH (Cognitive Orthosis for Assisting aCtivities in the Home) uses video monitoring and tracking technology to track behavior. COACH uses a partially observable Markov decision process to model the behavior and an artificial intelligence system coupled with audio and audio–video prompts to provide three levels of assistance depending on the person's level of need. In a pilot study using hand washing as the target activity, the COACH system showed promise in supporting independent functioning among participants with moderate dementia (Mihailidis, Boger, Craig, & Hoey, 2008). Other assistive technologies use digital television, cell phones (Donnelly, Nugent, Craig, Passmore, & Mulvenna, 2008), and handheld devices such as personal digital assistants (PDAs) (Becker & Webbe, 2008) to deliver reminders. Wayfinding technology that could prove useful to people with cognitive decline or dementia combines a global positioning system (GPS) with a wearable belt that delivers vibro-tactile signals to wearers to assist in navigation. This system could provide more independence to people in the early stages of AD, although early clinical trials suggest that as dementia progresses, individuals are less able to effectively interpret the vibro-tactile information (Grierson, Zelek, Lam, Black, & Carnahan, 2011).

12.2.5   Knowledge Transfer and Exploitation

There are many challenges facing the field of aging health technology. Many of the projects in the ETAC program have the potential to be transferred to the community at large but will likely require a cooperative approach among academics, clinical researchers, companies, software developers, and venture capital. It is not always easy for academics and clinicians to find the right way to move their prototype forward through a rigorous development and commercialization process. Some possibilities include developing relationships with business schools in one's own institution or to include MBA students in the development team or to participate in business concept competitions. These approaches can lead to different avenues and broader personal connections within the business world, thereby facilitating the introduction of a solution into the community at large. There is also a pressing need to understand and consider how these technologies will be marketed. Many, insofar as they are personal technologies, must address the issue of possible stigmatization associated with their use. This is a real problem that will impede adoption. The marketplace for assistive technologies is littered with products that failed to address this key issue. Usability is also a critical challenge in developing acceptance and ongoing use of these technologies. A thorough understanding of the human–computer interaction challenges inherent in a given solution, at the earliest stages of prototyping, will help form the basis and guidelines for more detailed design.

Beyond iterative prototyping and small deployments of these technologies is the need for robust evidence that not only addresses the clinical and academic questions but also provides guidance on key questions related to acceptance, adoption, and development. Attention to these aspects of solution development will also attract the attention of the various funding resources. Going forward, there is a need to study real people in real settings, to launch large-scale randomized controlled trials, to broadly disseminate the findings and implications, and to promote collaboration and cooperation among researchers in the field of aging health technology. There is tremendous opportunity for developing healthcare technologies at the moment but that opportunity can only be fully exploited by a comprehensive approach that brings all the stakeholders to the table as solutions are researched, developed, and put to use in the community at large.

12.3 Technology and Aging in the Asia-Pacific Region

The Asia-Pacific region (East and Southeast Asia) in mid-2010 was home to almost 2.2 billion people, just under one-third of the world's total population. In Southeast Asia, 6 % were aged 65+; in East Asia, the proportion was 10 %. If Australia and New Zealand are included, 13 % were in this age group. The percentage of older people varies considerably across the region, from 23 % in Japan aged 65+ to only 3 or 4 % in Brunei, Cambodia, Laos, Malaysia, the Philippines, and small Timor-Leste (Population Reference Bureau, 2010; UN ESCAP *Population Data Sheets*, annual). Moreover, many of the countries in the Asia-Pacific region are aging faster than the Western nations did, and many will double their older population in a compressed time span of 20–25 years (Phillips, Chan, & Cheng, 2010). The social, economic, and probably the technological effects of population aging will be felt most in the Asia-Pacific region in the coming years. This will stem from a combination of increased longevity in many countries, declining family size (from very reduced total fertility rates), and related impacts on the need for health and social care (Cheng, Chan, & Phillips, 2008; Kinsella & Phillips, 2005; Phillips et al., 2010).

The region has a variety of subregions in terms of propensity and penetration of various types of technology. There are countries with relatively large populations, though small percentages of older people, such as Indonesia, Cambodia, and the Philippines. Here, absolute levels of gerontechnology use and adoption are likely to be relatively low. At the other end, there are countries with high penetration and potential and considerable socioeconomically and culturally based demand, such as Japan, Korea, Taiwan, China, Hong Kong, Australia, and New Zealand. In between there are countries such as Thailand, Vietnam, and Malaysia, where considerable amounts of technology manufacture and use are to be seen and where this will ultimately involve older people. Indeed, many countries or parts of countries are the leading global locations for manufacture of many forms of technology (China, Vietnam, Thailand, etc.), and several are the global leaders in technological design and innovation, especially Japan, Korea, Taiwan, and increasingly China. This said, there are many issues with older populations regionally that can hinder technology uptake. Many countries have considerable proportions of their older populations, especially older females, who have limited literacy due to restricted education when they were young. Several of the languages are complex and have only been formulated for use in IT and keyboards relatively recently. Moreover, while numbers of older people in urban areas are substantial, with access to many forms of technology, even larger proportions of older people live in rural areas and still carry on agriculturally related activities and have little routine contact with technology.

As everywhere, technology issues tend to focus on either healthcare-related developments such as telemedicine, telecare, and health monitoring or assistive technologies for home and institutional living, in particular the potential for robotics and smart homes. There are, however, other themes in this region that are becoming important, as sales of technologically oriented consumer goods are increasingly seen as part of the growing *silver market* for older persons (Ong, Phillips, & Hamid, 2009). The growing affluence of many people, especially in the 50+ bracket, means older persons are consumers of audiovisual equipment, mobile phones, and personal computing equipment. Many items of health-related technology are made in the region, such as in-home blood pressure monitors, hearing aids, and various temperature measuring devices, and are becoming available at reasonable price.

Internet access is often regarded as an indicator of how *technology enabled* an area and population may be. In this respect, there is a huge range in the region. Japan, Korea, Australia, Taiwan, and Hong Kong have very high levels of Internet (and mobile phone) penetration in households, and it is assumed these technologies are accessible to older members. Japan, with the largest population of older persons, also has the largest percentage of Internet users among older adults—in 2006, 67 %—almost double the proportion in 2000 (McCreadie, 2010). Internet World Statistics (2010) shows Asia (as a whole) to have the largest number of Internet users—42 % of the global total—but behind the region's share of total population (approximately 55 %). Within the region, China has the largest number of users, 51 % of the region's total (420 million users, with a penetration of 31.6 %). Korea records the highest proportionate Internet usage statistics, with 40 million users (81.1 % penetration), followed by Japan with 99 million users (78.8 % penetration), then Singapore (77 %), Taiwan (70 %), Hong Kong (69 %), and Malaysia (65 %).

There are surprisingly high Internet penetration rates in many of the newly industrialized countries such as the Philippines (30 %), Vietnam (27 %), Thailand (26 %), and even Indonesia (12 %). However, there is a group of relatively Internet-isolated countries, such as Laos (7.5 %), Cambodia (0.5 %), and Timor-Leste (0.2 %). There is a great interest in communications technologies in the region, as evidenced by the annual APRICOT conferences (Asia Pacific Regional Internet Conference on Operational Technologies).

What does technology mean to older persons regionally? Examples are quite numerous in the literature, and a few will be cited. The first, the use of healthcare monitoring for older persons, involved international collaboration with a major pharmaceutical company. In this example, Tsang (2010) discusses tele-aging-care through a Tele Diabetes Monitoring and Management System in Hong Kong. He notes that telemedicine, the delivery of medicine at a distance, can involve monitoring a patient at home using well-known devices and transferring the information to a caregiver. Telemedicine may be simple, for example, two health professionals discussing a case over the telephone, or more complex, using satellite technology and videoconferencing equipment to conduct a real-time consultation between medical specialists in two different places. It may be real time (synchronous) and/or store and forward (asynchronous). Real-time telemedicine requires the presence of both parties at the same time and a communication link between them. Videoconferencing equipment is one of the most common technologies used in synchronous telemedicine. Store-and-forward telemedicine involves acquiring medical information and transmitting it for assessment off-line when convenient. Tsang also discusses a new emerging aspect of telemedicine, known as Primary Remote Diagnostic Visits, in which devices examine a patient. A doctor in another location virtually examines the patient and treats him or her. This hybrid technology holds great promise for solving healthcare delivery problems.

Other examples involve the development and application of smart home technology. Soar (2010) reports early findings of an Australian research program, the Queensland Smart Home Initiative. This research looks at the development and evaluation of demonstrator smart homes, hospital avoidance and connected communities for care, connectivity for information sharing in rural communities, policy, and strategies for assistive technology. Initial findings indicate many adverse events in which technology could assist, such as falls, memory loss, medication problems, and social isolation. Soar notes that while much innovative technology is available, levels of adoption remain low. Some progress is needed before providers such as health departments are ready to shift focus and resources toward prevention, assistive technology, and community care. There are also structural barriers, and as in many other places, the health system remains strongly dominated by the hospital sector.

From Korea, Bien (2010) discusses human–robot interaction (HRI) related to the various forms of service robotic systems in residential environments. If HRI is used to assist older people and people with disabilities, the total system operation needs to be human-friendly. To achieve such a human-friendly approach for older people in a smart home, the team adopted Computational Intelligence (aka soft computing technique) that effectively mimics human behavior and transplants soft knowledge to machines. The study reports on experiments undertaken in an *Intelligent Sweet*

Home in the Human-friendly Welfare Robot System Research Center at KAIST (Korea Advanced Institute of Science and Technology, located in the city of Daejeon, 150 km south of Seoul). The research used various Computational Intelligence techniques to try to achieve a human-friendly assistive robotic environment. A number of interaction–interfaces, including facial expression and hand gestures, were successfully implemented within a home setting. The study suggests that a living environment with a class of human-friendly service robotic systems can be achieved and independence of the residents can be facilitated through Computational Intelligence methods. Such a robotic service could become an important alternative for human care and assist older persons and people with disabilities.

Finally, Ng et al. (2010) have looked at the social impact of smart home technology in Singapore. In many Asian societies, older people are traditionally supported by their families and this may be viewed as a burden. Smart homes have the potential to improve the quality of life in older adults and help them live more independently, but little was known about Singaporean's attitudes toward the use of these technologies in their homes. Over the past 5 years, the research team at the Institute for Infocomm (I²R), A*STAR, in Singapore, has investigated the needs and attitudes of Singaporeans toward smart homes, using a range of survey methods. For example, a survey was conducted at the first Silver Industry Conference and Exhibition (SICEX, 2008), where a futuristic smart home was presented, to understand users' perception toward the featured technology. Respondents aged 55–90 in a day-care facility were surveyed. A smart home lab, STARhome, was also established and contextual interviews with three-generation families were conducted to understand their technology use and expectations of future homes. This wide range of studies revealed that many older adults own mobile phones and use them to communicate with their families and friends. In terms of household technology, well-known devices such as CD players, radios, washing machines, and TVs were more commonly owned and used. Some older adults own and use DVD players, but only a few knew how to operate computers. Most lead simple lives though some were more "tech-savvy" than others. The research enabled the team to conduct contextual interviews with some more tech-savvy older adults in their homes and the STARhome. Most respondents saw safety and security as the most immediate and obvious need to satisfy in a smart home, followed by health care. Interestingly, the team found the needs and attitudes of older adults differed from those of working adults and youth. The research also highlighted behavioral issues, such as medication compliance, and it was hoped that smart home technology would be able to help with reminding older adults to take medications.

12.4   Technology for Active Aging in the United Kingdom

The focus in this chapter so far has been on research and development (R&D) initiatives at an international and regional level of operation. Considerable activity is also funded and carried out within national programs and frameworks, and in this section the case of the United Kingdom is examined to outline the way R&D initiatives are linked to local circumstances and policy context. The UK is an interesting example on a number of counts. Firstly, it has a long history of use of technologies in the care of older people (Fisk, 2003), for instance, with the use of personal alarm systems linked to an emergency call center. Market penetration for these kinds of services is relatively high compared to most other countries. The organization of services for older people is also interesting, with responsibility for the support of older people shared between social services (local authorities, such as city councils) and health services (National Health Service general practitioners, community health services, acute services, etc.). While this can create problems of service integration, the emphasis on *social care* provides some counterpoint to a primarily medical model of care for older people. Moreover, since the late 1980s there has been a strong emphasis on home- and community-based services, rather than the delivery of care in institutional environments such as nursing homes.

In the UK, research and development activities concerned with the application of technology for active aging is being undertaken in many different ways and takes many forms, from commercial developments, some of which are highly speculative, through to basic research. There is a broad spectrum of work concerned with independent living, with a strong emphasis on Ambient Assisted Living and supporting mobility. Further, there is an emphasis on monitoring health conditions and the provision of care. Although much addresses the needs of those who are frail, the needs of active older people are an important driver of many developments, many of which involve European and international collaborations.

Many mainstream commercial activities have built upon the infrastructure that has supported the widespread use of the community alarm coupled with the opportunities provided by the increasing reliability and cost-effectiveness of wireless technologies, such that a multitude of smart home technologies are now available to support individuals with many different health conditions. However, much of the work, which extends beyond this into the telehealth and telecare arena, is at the stage of proof of concept and the evaluation of economic feasibility and commercial viability. Previously there were many mainly small-scale trials, but these resulted in largely ambiguous outcomes. Thus, a determined attempt is being made to undertake larger-scale trials, especially through Whole System Demonstrator projects (DH, 2010), in order to evaluate their effectiveness and to develop business cases. Further developments are being funded through the government-funded Technology Strategy Board's Assisted Living Innovation Platform (TSB, 2011). Significantly, this is actively supporting research on business and economic modeling and on social and behavioral studies. While the TSB's focus appears to be largely on technologies which are now well understood in terms of their potential to enhance active aging, independence, and quality of life, technology development extends much further (in universities and commercial settings) with important advances in sensor technologies which support the identification and management of chronic diseases.

Research and development in the UK is greatly enhanced by the government-funded Research Councils that support basic and applied research in universities and similar institutions but not commercial research. Through a highly competitive bidding process, individuals and teams of researchers can apply for funding support to specified programs of work. There have been a number of aging-related programs since 1997, some of which have supported work with a technology focus. The most important is the Engineering and Physical Sciences Research Council (EPSRC) Extending Quality Life (EQUAL) Initiative, which commenced in 1998 and is largely completed. This has been especially successful in supporting work to enhance the quality of life of older people with dementia and those suffering the effects of stroke and heart and lung disease, as well as developing technologies to enable better access in both urban and rural environments. In addition, EQUAL has supported design-orientated research which is leading to the design of homes, urban environments, and consumer products that are more inclusive of the needs of older adults and disabled people. Examples of EQUAL-funded projects include the following:

SMART (The SMART Consortium, 2011) being undertaken by a consortium of universities. The project explores how technology might be used to facilitate active in-home rehabilitation, initially by focusing on people suffering from the after effects of stroke and more recently on understanding the potential for technology in the support of self-management of conditions experienced by individuals with chronic pain and congestive heart failure. User-centered designs for technology, for example, resulting in a personalized self-management system, were a fundamental starting point for the work, so too were extensive studies with the technology in use. Users, carers, and clinical stroke therapists are an integral part of the SMART team, from initial scoping of need through to the testing of new iterations of the system as it develops. This is a significant study, especially since self-management is central to the UK Government's agenda for the management of long-term chronic conditions.

Introducing Assistive Technology into the Existing Homes of Older People (King's College London and the University of Reading, 2004). Through detailed analysis of the capability of many fixed and personal assistive technologies, this research addressed the feasibility and cost of adapting different types of domestic property to meet the needs of individuals with various impairments, some very severe. Through working with older users, it considered the acceptability of various assistive technologies and what leads to adoption or rejection. In particular it explored the relative costs of different care packages, which integrated various combinations of care services and provision of assistive technology. It showed the extent to which assistive technology could substitute and supplement conventional care services, the potential trade-offs between the two, and the likely returns on investment in housing adaptations and assistive technology.

Future Bathroom (Lab4Living, 2011) is an example of work that is helping to encourage a consumer perspective rather than an impairment perspective on assistive technology. This involves researchers, designers, and manufacturers in the quest for better quality and design of bathroom furniture for older people. The challenge is to design quality products that all bathroom users would find acceptable, as well as meeting the specific needs of older and disabled people. The project has developed a robust methodology for fostering codesign dialogues between designers, researchers, and older people (aged 50+) with chronic age-related health conditions such as arthritis, osteoporosis, stroke, and macular degeneration. The work is leading to a range of innovative and desirable bathroom prototypes (both bathroom furniture and assistive technologies for bathroom use) that are sensitive to the problems of living with disability, which do not stigmatize, are capable of manufacture, and demonstrate the principles developed as a result of the project.

Although EPSRC supports a number of significant technology projects, much of its funding for aging-related research is channeled through the New Dynamics of Aging (NDA, 2011) and lifelong health and well-being (LLHW) (MRC, 2011). Although these support some research projects with a technology orientation, in general technology is not prominent. There are some exceptions as well as a smattering of work concerned with the relationship of older people with technology. Examples of the work supported by NDA include the six projects described briefly below (NDA, 2011):

Design for Aging Well is considering the enhancement of autonomy and independence through the development of a clothing system with embedded wearable technologies. As with much current design research involving older and disabled people, it takes a user-centered design perspective, an approach that is rarely found in fashion design. Working with user groups, the project is developing and testing concepts for the system that older people will willingly wear.

Envision has evaluated an innovative way of communicating to health professionals the complexity of older people's mobility problems using visualizations of objective dynamic movement data. A prototype software tool visualizes, for non-biomechanical specialists and lay audiences, dynamic biomechanical data captured from older people undertaking ADLs. From motion capture data and muscle strength measurements, a 3D animated human *stick figure* is generated, on which the biomechanical demands of the activities are represented visually at the joints (represented as a percentage of maximum capability, using a continuous color gradient from green at 0 %, amber at 50 % through to red at 100 %). The potential healthcare and design applications for the visualizations have been evaluated with older people and healthcare and design professionals.

NANA is a 3-year multidisciplinary research project using sensitively designed technology to improve data collection and integrate information on nutrition, physical function, cognitive function, and mental health to identify individuals at risk of undernourishment and improve targeting of interventions. As well as improving the measurement, it will also enhance understanding of the interactions between these factors. This improved understanding may be used to inform strategies to prevent physical and mental decline in aging and to develop improvements in the medical treatment of older people. The resulting toolkit has commercial development potential primarily for use with older people and also with other groups that would benefit from comprehensive integrated assessment.

Sus-IT is exploring the relationship between the dynamics of aging and the dynamics of digital information and communication technologies (ICTs), in order to understand how ICTs can support or enrich quality of life and autonomy of older people as they age. It is known that digital ICTs have the potential to support older people to live independently, promote social inclusion, and facilitate access to commercial or government services. However, as people age they often experience decline in their physical and/or cognitive abilities which can make it difficult to continue use of, or keep up with, digital tools and services. The resulting disengagement from the digital world can constitute a significant reduction in quality of life for some older people.

While the potential of these examples of recent and current research projects to contribute to major developments in support of active aging is considerable, in reality, the path of research from development through to application is fraught with difficulties. Nevertheless, many of the projects are being executed with methodologies which recognize the value of user-centered approaches and co-design and which involve older people as experts rather than just subjects in the research process. There is a good track record of progress toward application, especially for those projects supported by EQUAL, and of influential developments, at the conceptual level of informing others about what is possible and how this might be achieved, not least policy makers, through to direct application in product development.

Support for the development and use of technology that promotes active aging comes from a variety of sources. Government-funded health and social services have a clear interest in evaluating and, where possible, adopting a range of technologies. These include traditional assistive technologies through to both sophisticated and simple telecare and telehealth systems, to enable older people to maintain their independence and lead active lives. However, some technologies are becoming mainstreamed, finding their way onto the main street (KT-EQUAL, 2009), and are becoming available in conventional consumer outlets rather than just specialist suppliers. Organizations in the charitable and voluntary sector, many of which are focused on particular health conditions and impairments, play an important role, providing information and advice to older people about which technologies are likely to be of benefit and how to access and use them. These organizations in particular are concerned to see the development of inexpensive and reliable technologies and, given the current widespread exclusion from the digital economy experienced by many people, they are particularly eager to see the development of technologies that are inclusive of all members of the population.

12.5   European Union Initiatives in Ambient Assisted Living

The EU has a long history of cross-national initiatives in R&D in the area of technologies for older and disabled people. For example, the EU's Technology Initiative for Disabled and Elderly People (TIDE) funded 55 projects between 1993 and 1997 to develop new technological tools and applications, with an aim to enhance autonomy and participation in social and economic life. The EU has a mandate to co-fund activities to stimulate R&D that will both develop new market opportunities for EU industry and at the same time achieve goals set out in social policy, for example, enhancing social participation of marginalized groups through an inclusive *information society*. This previous R&D effort has now culminated in the EU Ambient Assistive Living Joint Program (AAL-JP), which is a 5-year funded project (2008–2013) with a total budget of 700 million euros (EU, 2008). The project is jointly funded by public bodies, AAL partner states, the EU, and private industry. It forms part of the EU e-Inclusion agenda which aims to improve access and take-up of IT services among disadvantaged groups including the elderly (EU, 2007a). AAL-JP is motivated by a recognition of the increasing aging demographic in Europe and the opportunities this brings for citizens, health- and social care systems, and industry and is the first concrete activity to emerge out of the European Action Plan for "Aging Well in the Information Society" (EU, 2007b).

Ambient Assisted Living (AAL) is "…the use of information and communication technologies in a person's daily living and working environment to enable individuals to stay active longer, remain socially connected and live independently into old age" (EU, 2008, p. 2). Ambient Assisted Living combines ICTs, stand-alone assistive devices, and smart home technologies to help support older people to live independently within the community. AAL is designed to address multiple interrelated aims: to allow people to *age-in-place* by increasing their autonomy, to maintain self-confidence and mobility, to support health and functional capability, to promote active and healthy lifestyles, and to prevent social isolation (AALIANCE, 2009). Early technologies in this field included button-based alarm systems worn by the older person to allow them to raise an alarm should a problem requiring assistance arise. New ICT-based products and services have emerged, such as the use of sensors to detect potential emergency situations such as a fall or environmental hazard (e.g., flood or gas leak) and summon help without action on the user's part. However, the recent advances in ICTs have most potential for supporting the independence and health of older adults by extending the range and level of support to seniors living in the community. AAL-JP has identified the need for technology interventions that broadly address older people's well-being, autonomy, and social inclusion, while also supporting older people with functional impairments, and the treatment and prevention of challenging conditions such as chronic diseases. Specific aims include to (AALIANCE, 2012):

- *Extend the time people can live in their preferred environment by increasing their autonomy, self-confidence, and mobility*
- *Support maintaining health and functional capability of the elderly individuals*
- *Promote a better and healthier lifestyle for individuals at risk*
- *Enhance the security, to prevent social isolation and to support maintaining the multifunctional network around the individual*
- *Support carers, families, and care organizations, to increase the efficiency and productivity of used resources in the aging societies*

In addition to the need to meet the demands of an aging population, the AAL-JP funding represents an attempt to move toward marketable products and away from high cost technologies where there has been little market penetration. The design of gerontechnological interventions to date has been technology driven, and AAL-JP addresses this by stipulating the need for a multi-partner initiative that draws upon differing expertise and a multiuser perspective. The EU focus also recognizes that interventions have largely failed to address the needs of users at all stages of product development and hence have not been grounded in the needs and requirements of older people. As a result, a strong emphasis is placed on the application and development of technology that addresses usability and acceptability, by involving older people, informal carers, and other stakeholders at all stages of design, specification, and application.

The EU funds its R&D activities through the publication of periodic *calls*, which set down the overall aims and research themes, specific objectives, budgets, and research instruments (e.g., types of projects and collaborations). Despite this there are overall criteria that the funding proposals must address when submitting an application. This criterion promotes a pan-European approach to the research (minimum of three EU countries must be represented in the proposal), as well as a multidisciplinary partnership as all consortiums are required to involve one eligible business partner, SME partner, and end-user organization (up to a maximum of ten consortium partners). This promotes a multidisciplinary approach to the research and the recognition that AAL interventions require the contribution of a broad range of skills and expertise to ensure fit for purpose.

The AAL-JP *first thematic call* in 2008 addressed **well-being and care** and focused on ICT-based solutions for the prevention and management of chronic conditions of elderly people. The call identified the need for prevention, management, and support services that address the social and socioeconomic environment related to chronic conditions. A total of 23 projects were funded for 24–36 months in duration and varied from basic monitoring technologies that reacted to changes in well-being and health status to the implementation of interactive functional robotic devices that encouraged aging within the home environment. The projects directly addressed areas of self-management, encouraging independent living and the promotion of autonomy within the home, ensuring that older people feel supported within the home and continue to engage in activities that enable them to age in the preferred place of the home. The management of chronic conditions represents a considerable challenge and the program considers that AAL interventions can assist self-management and reduce the costs of care required for this group.

A *second thematic call* on **social interaction** was released in 2009 and focused on providing innovative ICT-based solutions aimed at helping people to be active and socially connected in the society as they age, from both a societal and personal perspective, effectively contributing to their health, to overall quality of life, and to social inclusion. Thirty projects were funded, each with a duration of between 18 and 36 months. The focus of the funded projects in phase 2 has been on facilitating the active contribution of the elderly, through projects that provide ICT solutions for older people to engage and sustain in social networks and to share information with each other as a process through which they can make a positive and active contribution. The funded projects incorporate a broad range of user-oriented services, including engagement in virtual environments and accessible mobile technologies connected via web-based servers so that older people can develop new social networks and feel a valued part of the community. The theme addresses wider issues relating to the e-Inclusion agenda, addressing social exclusion, isolation, and loneliness, and relating to the building of social capital among vulnerable groups who will increasingly depend on social networks for care and support.

A *third thematic call released* at the end of 2010 addressed **ICT-based solutions for Advancement of Older Persons Independence and Participation in the "Self Serve Society."** Although no details of funding or specific details of individual projects have been released, the theme considers AAL interventions that encourage personalized *self-serve* care, promoted through user choice and accessibility. This was to be achieved through projects that focus on preserving and enhancing independence and dignity in later life, active participation in the self-serve society, and ensuring that older people are provided with the support for participation. This fits with recent policy which has highlighted the need to reduce the costs of care delivery resulting from a globally aging population by putting the older person in control of accessing services. ICT-based interventions potentially allow services to be tailored to the needs of the older person while providing them with a gateway for accessing services.

The *fourth and final call* in 2011 addressed Advancement of Older Person's Mobility and aimed at the development of ICT-based solutions to help older persons to sustain their optimal level of mobility for as long as possible, as well as enhance a sense of confidence, autonomy, competence, security, and safety. The theme draws upon the understanding that mobility in later life encourages active aging, allowing older people to lead independent lives. This recognizes that AAL has an important role to play in helping older people to perform ADLs and engage in leisure activities which are important to delay the need for intensive care interventions. As well as the practical benefits, a good level of mobility promotes a sense of mental well-being, derived from a feeling of confidence when undertaking daily activities. Potential ICT solutions include supporting mobility in and outside the home through assistance with journey planning, monitoring and tracking movement, and assistive devices to support ambulation in the home.

The funding stream also responds to the debate concerning how AAL supports the needs of all stakeholders, not just older people, in providing interventions. The AAL program clearly responds to the desire for integrated service provision, identifying the need for technological interventions which recognize the needs of carers, families, and care organizations within an integrated system of care delivery. This considers AAL interventions as being part of the solution, where technology supports (and does not replace) other methods of administering care. In adopting an inclusive and integrated approach, the program makes reference to funding products, systems, or services that meet the *identified wishes and needs of end users*. However, the mechanisms for how user needs should be articulated in the application and development and AAL interventions are less well detailed yet indicate the drive toward a participatory ethos detailed in the broader e-Inclusion agenda of which the AAL-JP program is a part. The projects within the funding body require evidence-based user-centered research, incorporating user involvement in the application and development of many of the initiatives.

The AAL-JP program is also focused upon the development of tangible products, requiring that each project delivers a 2–3-year time-to-market perspective after the project ends. Given the short turnaround, project activity has involved the application and testing of technology which is already at the stage of specification and development, as opposed to funding work collecting new requirements. The program identifies the need to develop competitive industries, through the involvement of innovative companies, mostly small- and medium-sized enterprises (SMEs), to emerge from the niche markets of expensive devices (where much of the contribution has been thus far) to affordable devices for the mass market. Moreover, the tender documents identify the requirement for products that represent cost savings, in health cost, social costs, and nursing homes expenses, identifying the deficit that currently exists in many economies in meeting the care burden.

12.6 Discussion

The above scan of the international scene in the area of technology and aging is useful on a number of counts, even though it represents only a partial view of a rapidly growing area of research. First of all, researchers are often surprisingly blinkered with their perspectives focused on their own national or local contexts. The need for collaborative research, increasingly at an international level, is something that is very much part of the research scene in this area of work. Indeed, the mix of skills and expertise needed makes it impossible to visualize research in this field that does not cross the traditional disciplinary boundaries. The parallel global processes of population aging and technological change and development are also factors that are serving to drive local and national research and policy agendas together. For example, the UK national-level initiatives described above have been very active in fostering international collaborations, such as UK–Canadian joint research and networking. Second, it is useful to attempt to provide a summary of the state of the art in the area, even though reviews of technology-related initiatives and projects are almost by definition going to be immediately out of date. The underlying aim of this chapter has been to point to commonalities across programs and geographical regions, and this final section of the chapter aims to point to some of the key themes and directions in the research and identify lessons learned, or indeed lessons ignored, from these various initiatives.

12.6.1 The Theme of Active Aging

One of the striking aspects of the reviews is the wide range of applications that are under development. Academic research and commercial products in technology have historically focused on systems and devices that have had a strong medical component (telehealth) or have addressed issues of safety and security (telecare) (Sixsmith, 2006). In respect to telecare, it is possible to characterize technologies in terms of *three generations* of technology. The first generation refers to the simple community alarms that allow people to push a button on a telephone or pendant to raise an alarm if they had a fall or experience some other kind of emergency. The second generation is currently emerging into the marketplace and it comprises smarter devices that use sensors to check on safety of devices (e.g., cookers, baths) or to check on a person's level of activity. A third generation is in the R&D phase and aims to exploit recent advances in ICTs. The potential of these technologies to contribute in many areas of the everyday life of seniors is reflected in the much broader scope of the research. For example, both the ETAC and EU programs have a strong social agenda, driving research in systems to facilitate social participation. A further common feature is a focus on systems to help individuals, carers, and professionals to manage chronic disease. These include physical conditions, such as respiratory disease, but most notably include a growing number of research projects on helping people with dementia. This is a comparatively recent trend; as Sixsmith (2006) in a review of research programs prior to 2005 noted, there had been relatively little attention given to this group up to that time. Generally speaking the trend in research is toward an agenda that emphasizes active aging, rather than the agenda of dependency that has typified care of older people over the years.

12.6.2 Knowledge Transfer and Exploitation

Another emerging theme from the global scan is the high priority given to knowledge transfer (KT) and the exploitation of research results. Indeed, a common criticism of much of the research on technology and aging to date has been that there has been little payback in terms of real-world products and services for older people (Meyer, Muller, & Kubitschke, 2012). While there remain significant technical challenges, it is important that R&D in the field becomes more aligned with the nontechnical challenges and barriers to exploitation. A key element in this is the increasing emphasis on participatory approaches to research that includes stakeholders not only as research subjects but also as active participants in all stages of the R&D cycle. In addition to their traditional roles in field trials and evaluations, participants are now involved in the setting of research agendas, requirements specification, data collection, analysis, and interpretation. The involvement of older people, caregivers, and service providers is much more likely to result in products and services that are grounded in real-world needs and contexts, thus enhancing the likelihood of adoption. For example, the discussion of the ETAC program highlighted some of the potential downsides of technology, notably the potential for stigmatizing the very people it is intended to help. The challenge to the research community is to find appropriate ways of facilitating this active participation through innovative methods and to go beyond the often notional concepts of stakeholder involvement that currently persist. Examples of this include visualization techniques (see Chap. 9) that allow people to contribute to decision-making in virtual space prior to implementation of design solutions in the real world.

This shift in focus to real-world products and services is reflected in the various geographical reviews in this chapter. The discussion of the ETAC program highlighted the very practical problems and possible solutions regarding how people working in the academic sector can move their ideas and prototypes through the very challenging process of commercialization. An emphasis on commercialization is apparent in the UK and EU technology programs, emphasizing academic/commercial partnerships and the explicit inclusion of commercial objectives in the commissioning and funding of new research. For example, EU projects funded under the EU's AAL-JP are expected to result in near-to-market systems and devices. The extent to which these expectations are actually realized remains to be seen. The discussion of the Asia-Pacific region highlights a different dimension: the emergence of the old-age demographic in its own right as a market for new technology. This has been a facet of technologies-for-aging for many years, for example, the development and use of emergency alarms emerged in the 1960s through development and commercialization within the private sector. However, the rapidly aging populations worldwide and increasing familiarity of many older people with ICT-based products and services may see new everyday products and services emerge through entirely commercial channels in parallel to the more academic route. (Examples of these include some of the games that have recently come on to the market, such as *Brain Age.*) Despite this last point about the organic emergence of markets, the role of governments in promoting the use of ICT-based solutions for older people appears to remain an important factor. This is very apparent in the multibillion euro investment made by the EU through the AAL-JP, a program that is driven by expectations of wider economic and commercial benefits, as well as benefits to older users themselves. The discussion of the UK highlights the very active role that the government in that country has taken in deploying technology within the range of health and social services available to older people. Additionally, the UK is providing investment into these emerging services and considerable emphasis has been placed on developing a strong evidence base for policy and practice. The Whole System Demonstrator has been an important step, not only for evaluating health outcomes and cost savings but also for providing clear pathways for the scaling-up of emerging technology-based systems into mainstream services (DH, 2011).

12.6.3 Global Diversity

Even though there are strong global processes of population aging, economic mechanisms, institutions, and technological change that are serving to bind together nations and geographical regions in terms of common experiences and challenges, there remains considerable diversity within the world. This point is particularly apparent in the discussion of the Asia-Pacific region. Gerontological research tends to emphasize a Western perspective, yet in comparative terms, the so-called *problem of aging* is perhaps less challenging in North America and Europe compared to the situation in the Asia-Pacific region. The very rapid pace of population aging and the sheer numbers of older people in combination with cultural, social, and economic change set the Asia-Pacific region apart. This region, however, is also characterized by huge diversity internally, particularly within the older populations who in many of the East Asian countries have literally been left behind by rapid technological and social change and urbanization. While this chapter has pointed to common problems and common solutions (a good example is the need for better medication compliance and the potential of technologies to facilitate this), the international R&D collaborations and the products and systems that are beginning to emerge into the global market place need to be cognizant of the need to apply these solutions in a way that reflects diversity. The term *glocalization*, the adapting of global-level products, services, policies, and practice to reflect local needs and contexts, has to become increasingly a part of the KT lexicon in the area of technology and aging.

References

AALIANCE. (2009). *Ambient assisted living roadmap.* Retrieved from http://www.aaliance.eu/public/documents

AALIANCE. (2012). Retrieved from http://www.aal-europe.eu/about-us

Bas, E., Erdogmus, D., Ozertem, U., & Pavel, M. (2008). Towards fish-eye camera based in-home activity assessment. *Conference Proceedings – IEEE Engineering in Medicine and Biology Society, 2008,* 2558–2561.

Becker, S. A., & Webbe, F. M. (2008). The potential of hand-held assistive technology to improve safety for elder adults aging in place. In K. Henriksen, J. Battles, M. Keyes, & M. Grady (Eds.), *Advances in patient safety: New directions and alternative approaches* (Technology and Medication Safety, Vol. 4). Rockville, MD: Agency for Healthcare Research and Quality.

Bien, Z. Z. (2010). Human-friendly system approach in a robotic smart home for older persons. *Gerontechnology, 9*(2), 162.

Cheng, S. T., Chan, A. C. M., & Phillips, D. R. (2008). Ageing trends in Asia and the Pacific. In UN Department of Economic and Social Affairs (Ed.), *Regional dimensions of the ageing situation* (pp. 35–69). New York: United Nations.

Dale, C. L., Simpson, G. V., Foxe, J. J., Luks, T. L., & Worden, M. S. (2008). ERP correlates of anticipatory attention: Spatial and non-spatial specificity and relation to subsequent selective attention. *Experimental Brain Research, 188,* 45–62.

DH. (2010). *Whole systems demonstrators: An overview of telecare and telehealth.* Accessed May 19, 2011 from http://www.dh.gov.uk/prod_consum_dh/groups/dh_digitalassets/documents/digitalasset/dh_100947.pdf

DH. (2011). *Whole Systems Demonstrator Project: Headline figures.* Retrieved from http://www.dh.gov.uk/en/Publicationsandstatistics/Publications/PublicationsPolicyAndGuidance/DH_131684

Donnelly, M. P., Nugent, C. D., Craig, D., Passmore, P., & Mulvenna, M. (2008). Development of a cell phone-based video streaming system for persons with early stage Alzheimer's disease. *Conference Proceedings – IEEE Engineering in Medicine and Biology Society, 2008,* 5330–5333.

EU. (2007a). *European i2010 initiative on e-Inclusion* (Published on the 8th November 2007). Accessed January 6, 2012 from http://ec.europa.eu/information_society/activities/einclusion/docs/i2010_initiative/comm_native_com_2007_0694_f_en_acte.pdf

EU. (2007b). *European i2010 initiative on ageing well in the information society* (Published on the 14th June 2007). Accessed January 6, 2012 from http://eur-lex.europa.eu/LexUriServ/LexUriServ.do?uri=COM:2007:0332:FIN:EN:PDF

EU. (2008). *Overview of the ambient assisted living programme.* Accessed January 6, 2012 from http://ec.europa.eu/information_society/activities/einclusion/docs/ageing/aal_overview.pdf

Fisk, M. (2003). *Social alarms to telecare: Older people's services in transition.* Bristol: Policy.

Grierson, L. E. M., Zelek, J., Lam, I., Black, S. E., & Carnahan, H. (2011). Application of a tactile way-finding device to facilitate navigation in persons with dementia. *Assistive Technology, 23*(2), 108–115.

Hayes, T. L., Abendroth, F., Adami, A., Pavel, M., Zitzelberger, T. A., & Kaye, J. A. (2008). Unobtrusive assessment of activity patterns associated with mild cognitive impairment. *Alzheimers Dementia, 4,* 395–405.

Internet World Statistics. (2010). *World internet users and population stats.* Accessed March 21, 2011 from http://www.internetworldstats.com/stats.htm

King's College London and the University of Reading. (2004). *Introducing assistive technology into the existing homes of older people: Feasibility, acceptability, costs and outcomes.* London: Institute of Gerontology King's College London. ISBN 1-872342-17-5.

Kinsella, K., & Phillips, D. R. (2005). Global aging: The challenge of success. *Population Bulletin, 60*(1), 1–40 (available on http://www.prb.org)

KT-EQUAL. (2009). *Assistive technologies: What is their place in the mainstream market?* Accessed May 19, 2011 from http://kt-equal.org.uk/uploads/monographs/assistive_technologies_monograph.pdf

Lab4Living. (2011). *Future bathroom.* Accessed May 19, 2011 from http://www.lab4living.org.uk/future-bathroom

Mahoney, D. F., Mahoney, E. L., & Liss, E. (2009). AT EASE: Automated Technology for Elder Assessment, Safety, and Environmental monitoring. *Gerontechnology, 8*(1), 11–25.

Massimi, M., Berry, E., Browne, G., Smyth, G., Watson, P., & Baecker, R. M. (2008). An exploratory case study of the impact of ambient biographical displays on identity in a patient with Alzheimer's disease. *Neuropsychological Rehabilitation, 18*(5–6), 742–765.

McCreadie, C. (2010). Technology and older people. In C. Phillipson & D. Dannefer (Eds.), *The handbook of gerontology* (pp. 607–617). London: Sage.

Meyer, I., Muller, S., & Kubitschke, L. (2012). AAL markets—knowing them, reaching them: Evidence from European research. In Augusto et al. (Eds.), *Handbook of ambient assisted living.* Amsterdam: IOS Press.

Mihailidis, A., Boger, J., Craig, T., & Hoey, J. (2008). The COACH prompting system to assist older adults with dementia through handwashing: An efficacy study. *BMC Geriatrics, 7,* 8–28.

MRC. (2011). *Life long health and wellbeing.* Accessed May 19, 2011 from http://www.mrc.ac.uk/Ourresearch/ResearchInitiatives/LLHW/index.htm

NDA. (2011). *New dynamics of ageing.* Accessed 19 May, 2011 from http://www.newdynamics.group.shef.ac.uk/

Ng, J. S.-L., Tan, O. Y.-L., Koh, W.-K., Tay, Z. Y.-C., Wong, A. H.-Y., & Helander, M. G. (2010). Needs and attitudes of older adults and their families in Singapore towards smart homes. *Gerontechnology, 9*(2), 162.

Ong, F.-S., Phillips, D. R., & Hamid, T. A. (2009). Ageing in Malaysia: Progress and prospects. In T. H. Fu & R. Hughes (Eds.), *Ageing in East Asia: Challenges and policies for the twenty-first century* (pp. 138–160). London: Routledge.

Parsons, T. D., Silva, M., Pair, J., & Rizzo, A. A. (2008). Virtual environment for assessment of neurocognitive functioning: Virtual reality cognitive performance assessment test. *Studies in Health Technology and Informatics, 132,* 351–356.

Phillips, D. R., Chan, A. C. M., & Cheng, S.-T. (2010). Ageing in a global context: The Asia-Pacific region. In C. Phillipson & D. Dannefer (Eds.), *The handbook of gerontology* (pp. 430–446). London: Sage.

Population Reference Bureau. (2010). *2010 World population data sheet.* Washington, DC: PRB.

SICEX. (2008). *Convention & exhibition to highlight potential of billion dollar silver-hair market.* Retrieved from http://app1.mcys.gov.sg/PressRoom/SICEX2008ConventionExhibitionto Highlight.aspx

Sixsmith, A. (2006). New technologies to support independent living and quality of life for people with dementia. *Alzheimer's Care Quarterly, 7*(3), 194–202.

Soar, J. (2010). Smart home and assistive technology developments in Australia. *Gerontechnology, 9*(2), 161.

Sorrell, J. A., & Sorrell, J. M. (2008). Music as a healing art for older adults. *Journal of Psychosocial Nursing and Mental Health Services, 46,* 21–24.

The SMART Consortium. (2011). Accessed May 19, 2011 from http://www.thesmartconsortium. org/

Tsang, P. (2010) Tele ageing care: A case study of tele diabetes management system (TDMMS), *Gerontechnology,* 9 (2), 160–161

TSB. (2011). *Assisted living.* Accessed from http://www.innovateuk.org/ourstrategy/innovation-platforms/assistedliving.ashx

Tyrer, H. W., Alwan, M., Demiris, G., He, Z., Keller, J., Skubic, M., et al. (2006). Technology for successful aging. *Conference Proceedings – IEEE Engineering in Medicine and Biology Society, 1,* 3290–3293.

Tyrer, H. N., Aud, M. A., Alexander, G., Skubic, M., & Rantz, M. (2007). Early detection of health changes in older adults. *Conference Proceedings – IEEE Engineering in Medicine and Biology Society, 2007,* 4045–4048.

Index

Printed in the United States
By Bookmasters